New Frontiers in Materials Design for Laser Additive Manufacturing

New Frontiers in Materials Design for Laser Additive Manufacturing

Editors

Bilal Gökce
Eric Jägle
Manfred Schmid

MDPI • Basel • Beijing • Wuhan • Barcelona • Belgrade • Manchester • Tokyo • Cluj • Tianjin

MDPI

Editors
Bilal Gökce
University of Wuppertal
Germany

Eric Jägle
Bundeswehr University
Munich
Germany

Manfred Schmid
Innovation Center for
Additive Manufacturing
Switzerland

Editorial Office
MDPI
St. Alban-Anlage 66
4052 Basel, Switzerland

This is a reprint of articles from the Special Issue published online in the open access journal *Materials* (ISSN 1996-1944) (available at: https://www.mdpi.com/journal/materials/special_issues/mat-des-LAM).

For citation purposes, cite each article independently as indicated on the article page online and as indicated below:

LastName, A.A.; LastName, B.B.; LastName, C.C. Article Title. *Journal Name* **Year**, *Volume Number*, Page Range.

ISBN 978-3-0365-5881-3 (Hbk)
ISBN 978-3-0365-5882-0 (PDF)

Cover image courtesy of Bilal Gökce

Contents

About the Editors

Bilal Gökce

Bilal Gökce started his academic career at RWTH Aachen University and Fraunhofer ILT, where he completed his diploma in physics in 2008. He pursued his Ph.D. in solid state physics at North Carolina State University, USA where he was granted the Provost's Fellowship in 2009. He received his doctorate in 2012 and continued his career path as a post-doctoral researcher on non-linear optics. After moving back to Germany, he worked as a researcher at T-Systems International on laser applications. He has dedicated his later career to academia and was appointed as a group leader for Laser Material Processing & Additive Manufacturing at the University of Duisburg Essen in 2014. In 2018, he received his habilitation and Venia Legendi in Technical Chemistry at the University of Duisburg Essen. In the same year, he was awarded the Fojtik-Henglein Prize for his pioneering research results in the field of the laser synthesis of colloids. In 2020, he was accepted into the DFG Heisenberg Program. He is currently a full professor of Materials Science and Additive Manufacturing at the University of Wuppertal, holding one of the few Heisenberg professorships funded by the German research foundation. His research is driven by the goal of harnessing the outstanding properties of functional nanoparticles in powder materials, demonstrators and products for additive manufacturing.

Eric Jägle

Eric Jägle studied materials science at the University of Stuttgart, Germany, and at the University of Cambridge, UK, (2001–2007). For his Ph.D. (2011), he worked at the Max-Planck-Institut für Metallforschung (MPI for Metals Research) under the supervision of Prof. E. J. Mittemeijer. His work focused on the mesoscopic simulation of microstructure development during phase transformations, in particular during phases of recrystallization. Afterwards, he moved to the Max-Planck-Institut für Eisenforschung (MPI for Iron Research) in Düsseldorf, Germany. There, he worked as post-doctoral researcher in the department of Prof. D. Raabe on atom probe tomography analysis of electrical steels, precipitation transformations and mechanical alloying. In 2015, he became the leader of a newly formed group in the same department, working on alloys for additive manufacturing. In 2020, he moved to the Institute of Materials Science of the Bundeswehr University Munich as a full professor. His research focuses on various aspects of alloys used in AM such as particle reinforcement, in-process strengthening reactions, hot cracking behavior, residual stress and in-process metal–gas reactions. The investigated materials include steels, Ni- and Al- based alloys and composites.

Manfred Schmid

Manfred Schmid began his professional career with an apprenticeship as a chemical laboratory assistant at Metzeler Kautschuk AG in Munich. After graduating from high school on the second educational path, he studied chemistry at the University of Bayreuth, followed by a doctorate in the field of macromolecular chemistry. After graduation, he moved to Switzerland for professional reasons, and 17 years followed with various industrial positions in the field of polymer research and production. There, he also worked with materials testing for engineering thermoplastics and in polymer analysis. Polyamides and biopolymers were the primary focus of the various activities. For around thirteen years, he has headed the research area for laser sintering (LS) at Inspire AG. Inspire AG is the Swiss competence center for production technology. It acts as a transfer institute between universities and Swiss industry. The focus of his current work is on new polymers for the LS process, analytical evaluation of LS powders with respect to their specific property profiles, and LS process development. He supervises several collaborators and has led numerous of different research projects in this field. This has resulted in a number of frequently cited original publications.

materials

MDPI

Editorial

New Frontiers in Materials Design for Laser Additive Manufacturing

Silja-Katharina Rittinghaus [1], Eric A. Jägle [2], Manfred Schmid [3] and Bilal Gökce [1,*]

1 Chair of Materials Science and Additive Manufacturing, School of Mechanical Engineering and Safety Engineering, University of Wuppertal, 42119 Wuppertal, Germany
2 Institute of Materials Science, Universität der Bundeswehr München, 85579 Neubiberg, Germany
3 Innovation Center for Additive Manufacturing, Inspire AG, 9014 St. Gallen, Switzerland
* Correspondence: goekce@uni-wuppertal.de

Laser-based additive manufacturing (LAM) in all its variations is now being established as a technique for manufacturing components from various material types and alloys. However, the materials regularly used in these processes were developed for conventional manufacturing processes (e.g., casting, injection molding, thermal spraying). They are therefore not optimized for the characteristic process environments in laser additive manufacturing, so some of the great potentials of this manufacturing technology remain untapped. The urgent need for new materials throughout the industry is reflected in current trends and market studies, e.g., [1], as well as intensified research activities on material development for (L)AM.

The task of designing new materials is very challenging and, to be truly successful, requires interdisciplinary collaboration between experts from a wide range of disciplines (cf. Figure 1). Promising approaches to research include additivation and modification of existing commercial base materials, typically powders, but also creating completely new alloys starting with phase modeling and basic chemical reactions. Special properties to be taken into account are, for example, powder material properties, flowability properties, melt pool and flow characteristics, as well as solidification conditions. The consideration of industrial requirements such as high efficiency, reproducibility, and precision is highly essential for a holistic, sustainable approach.

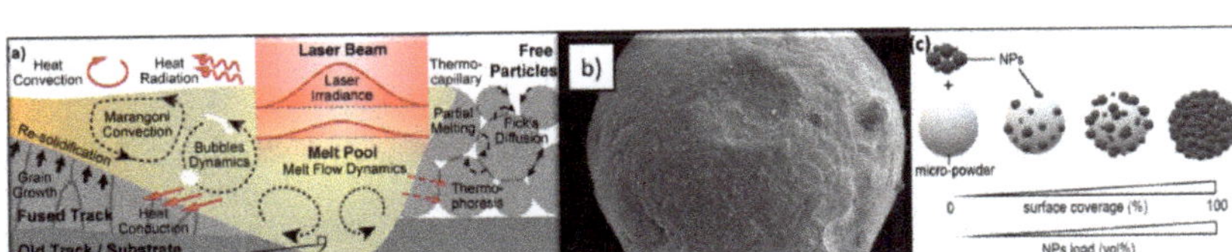

Figure 1. (**a**) Schematic of interactive physical phenomena during a stable LPBF process [2]. (**b**) scanning electron microscopy (SEM) picture of 1.2709 at 10,000× *g* magnification coated with 1 vol.% SiC [3]. (**c**) Illustration of the surface coverage by micro powders with increasing nanoparticle loading (vol%) [4].

With regard to metallic materials, one long-lasting challenge is to modify previously un- or hardly processable materials in such a way that their defect-free consolidation becomes possible, e.g., by adding zirconia or boride nanoparticles to highly crack-susceptible alloys such as Al7075 and Al606, as it is successfully being pursued by [5], and thus to expand the application range of LAM. This goes hand in hand with research and development on the processes themselves, as, e.g., proposed by [6], accepting the challenge of manufacturing Mg-alloys with laser-based powder bed fusion (LPBF). Further developments and innovations in the fields of systems engineering and photonics support these efforts, such as

Citation: Rittinghaus, S.-K.; Jägle, E.A.; Schmid, M.; Gökce, B. New Frontiers in Materials Design for Laser Additive Manufacturing. *Materials* **2022**, *15*, 6172. https://doi.org/10.3390/ma15176172

Received: 26 August 2022
Accepted: 1 September 2022
Published: 5 September 2022

the use of ultrasound [7] or green laser radiation [8]. Similarly, many studies are currently in progress to determine how the special temperature conditions in the process (cyclic heating, rapid cooling) can be used in a targeted manner to produce special microstructures, for example, precipitates or extremely fine structures, like in Al-Ni eutectic alloys investigated by [9] or even to adjust properties locally in the component, as can be derived from results of [10] on Al-Fe. Intrinsic heat effects can be used to trigger material transformations if they are cleverly controlled [11] and thus enable new component designs. Likewise, a targeted scan strategy and process parameter control allows us to locally promote and select crystallographic orientations and phase compositions (e.g., [12–14]).

Due to the increasing computational capacity and performance, numerical methods are increasingly used to represent and predict material behavior. Modeling of non-equilibrium states, such as those encountered in LAM, is a valuable complement to classical phase diagrams and paves the way for digital material development. Additivation, primarily with nano-sized additional particles that are not dissolved in the melt pool and act as nuclei and/or reinforcement in the resulting microstructure, is another growing field of research for LAM of metals. Oxide dispersion strengthening (ODS) is an evolving example of how material development benefits from AM by providing a method to produce these types of composites more economically or at all. These advances also benefit from sophisticated modeling, e.g., [2] propose a model to predict the nanoparticle's exact location in solidified LAM material. Another promising approach to modify powders at the nano level is to systematically coat them to improve process behavior, like, e.g., investigated by [3] on maraging tool steel.

Nanoparticles are also quite exciting for the AM of polymer materials [15]. A lot of research focuses on semi-crystalline polyamide 12 (PA12). The authors of [16] were able to prove in their investigations that even small amounts of added carbon nanoparticles can significantly influence the mechanical properties without adversely changing the crystallization behavior. Metallic or ceramic additives in micro- or nanoscale embedded in a polymer matrix can be used to produce highly specialized parts even with new functionalities, like, e.g., magnetic characteristics [17]. Achieving the desired distribution, however, is still challenging in many applications. Polylactide acid (PLA) is another polymer with increasing popularity in AM. Known to be both biocompatible and biodegradable, the scope in [18] to also make the manufacturing route more environmentally friendly is both obvious and ambitious.

In both metal and plastic AM, sustainability is playing an increasingly important role in material selection and development. Efforts in the metals sector often include attempts to substitute elements that are environmentally and ethically critical or that have an insecure supply chain, such as rare earths, and to improve recycling routes for materials along the entire process chain and components. With regard to polymers, research into the more ecological production of raw materials is coming to the fore, as well as investigations into materials that are biodegradable or easier to recycle or reuse.

Another material class-independent trend is the increasing use of modeling to map process-dependent material behavior, thus reducing experimental effort and shortening development cycles. For this, many correlations remain to be fully understood, making fundamental work like, e.g., [19], which focuses on capillary phenomena, essential. This is also accompanied by ever-increasing demands on measurement technology, which motivates work like, e.g., those of [20], which contributes to learning about fast phase transition kinetics by using advanced measurement techniques. In general, in situ techniques for observing microstructure development during the process to gain a deeper knowledge of process–property correlations are on the rise and quite successful [21].

As in any field of research, a collaboration between different researchers is the key to innovation. Increasing digitization has advanced the rapid availability of data worldwide and is accelerating developments on a massive scale. To assure consistency and validate developed methods, interlaboratory studies like the one proposed by [4] are useful evaluation tools that benefit from a high number of researchers and laboratories involved.

Additionally, it becomes increasingly crucial to keep track of the growing volumes of data and make the best possible use of them, research data management is also gaining more and more importance.

Conflicts of Interest: The authors declare no conflict of interest.

References

1. Wohlers, T.T.; Campbell, R.I.; Diegel, O.; Kowen, J.; Huff, R.; Mostow, N. *Wohlers Report 2021*; Wohlers Associates: Fort Collins, CO, USA, 2021; ISBN 978-0-9913332-7-1.
2. Yang, Y.; Doñate-Buendía, C.; Oyedeji, T.D.; Gökce, B.; Xu, B.-X. Nanoparticle Tracing during Laser Powder Bed Fusion of Oxide Dispersion Strengthened Steels. *Materials* **2021**, *14*, 3463. [CrossRef] [PubMed]
3. Pannitz, O.; Großwendt, F.; Lüddecke, A.; Kwade, A.; Röttger, A.; Sehrt, J.T. Improved Process Efficiency in Laser-Based Powder Bed Fusion of Nanoparticle Coated Maraging Tool Steel Powder. *Materials* **2021**, *14*, 3465. [CrossRef] [PubMed]
4. Kusoglu, I.M.; Huber, F.; Doñate-Buendía, C.; Rosa Ziefuss, A.; Gökce, B.; Sehrt, J.T.; Kwade, A.; Schmidt, M.; Barcikowski, S. Nanoparticle Additivation Effects on Laser Powder Bed Fusion of Metals and Polymers—A Theoretical Concept for an Inter-Laboratory Study Design All Along the Process Chain, Including Research Data Management. *Materials* **2021**, *14*, 4892. [CrossRef] [PubMed]
5. Martin, J.H.; Yahata, B.D.; Hundley, J.M.; Mayer, J.A.; Schaedler, T.A.; Pollock, T.M. 3D printing of high-strength aluminium alloys. *Nature* **2017**, *549*, 365–369. [CrossRef] [PubMed]
6. Julmi, S.; Abel, A.; Gerdes, N.; Hoff, C.; Hermsdorf, J.; Overmeyer, L.; Klose, C.; Maier, H.J. Development of a Laser Powder Bed Fusion Process Tailored for the Additive Manufacturing of High-Quality Components Made of the Commercial Magnesium Alloy WE43. *Materials* **2021**, *14*, 887. [CrossRef] [PubMed]
7. Todaro, C.J.; Easton, M.A.; Qiu, D.; Zhang, D.; Bermingham, M.J.; Lui, E.W.; Brandt, M.; StJohn, D.H.; Qian, M. Grain structure control during metal 3D printing by high-intensity ultrasound. *Nat. Commun.* **2020**, *11*, 142. [CrossRef] [PubMed]
8. Gruber, S.; Stepien, L.; López, E.; Brueckner, F.; Leyens, C. Physical and Geometrical Properties of Additively Manufactured Pure Copper Samples Using a Green Laser Source. *Materials* **2021**, *14*, 3642. [CrossRef] [PubMed]
9. Boussinot, G.; Döring, M.; Hemes, S.; Schmidt, M. Laser powder bed fusion of eutectic Al–Ni alloys: Experimental and phase-field studies. *Mater. Des.* **2016**, *198*, 109299. [CrossRef]
10. Wang, W.; Takata, N.; Suzuki, A.; Kobashi, M.; Kato, M. Microstructural Variations in Laser Powder Bed Fused Al–15%Fe Alloy at Intermediate Temperatures. *Materials* **2022**, *15*, 4497. [CrossRef] [PubMed]
11. Kürnsteiner, P.; Wilms, M.B.; Weisheit, A.; Gault, B.; Jägle, E.A.; Raabe, D. High-strength Damascus steel by additive manufacturing. *Nature* **2020**, *582*, 515. [CrossRef] [PubMed]
12. Sofinowski, K.A.; Raman, S.; Wang, X.; Gaskey, B.; Seita, M. Layer-wise engineering of grain orientation (LEGO) in laser powder bed fusion of stainless steel 316L. *Addit. Manuf.* **2021**, *38*, 101809. [CrossRef]
13. Kim, D.; Ferretto, I.; Jeon, J.B.; Leinenbach, C.; Lee, W. Formation of metastable bcc-δ phase and its transformation to fcc-γ in laser powder bed fusion of Fe–Mn–Si shape memory alloy. *J. Mater. Res. Technol.* **2021**, *14*, 2782–2788. [CrossRef]
14. Sun, Z.; Tan, X.; Tor, S.B.; Chua, C.K. Simultaneously enhanced strength and ductility for 3D-printed stainless steel 316L by selective laser melting. *NPG Asia Mater.* **2018**, *10*, 127–136. [CrossRef]
15. Kusoglu, I.M.; Doñate-Buendia, C.; Barcikowski, S.; Gökce, B. Laser Powder Bed Fusion of Polymers: Quantitative Research Direction Indices. *Materials* **2021**, *14*, 1169. [CrossRef] [PubMed]
16. Sommereyns, A.; Gann, S.; Schmidt, J.; Chehreh, A.B.; Lüddecke, A.; Walther, F.; Gökce, B.; Barcikowski, S.; Schmidt, M. Quality over Quantity: How Different Dispersion Qualities of Minute Amounts of Nano-Additives Affect Material Properties in Powder Bed Fusion of Polyamide 12. *Materials* **2021**, *14*, 5322. [CrossRef] [PubMed]
17. Hupfeld, T.; Salamon, S.; Landers, J.; Sommereyns, A.; Doñate-Buendía, C.; Schmidt, J.; Wende, H.; Schmidt, M.; Barcikowski, S.; Gökce, B. 3D Printing of magnetic parts by Laser Powder Bed Fusion of iron oxide nanoparticle functionalized polyamide. *J. Mater. Chem. C* **2020**, *8*, 12204–12217. [CrossRef]
18. Dechet, M.A.; Demina, A.; Römling, L.; Gómez Bonilla, J.S.; Lanyi, F.J.; Schubert, D.W.; Bück, A.; Peukert, W.; Schmidt, J. Development of poly(L-lactide) (PLLA) microspheres precipitated from triacetin for application in powder bed fusion of polymers. *Addit. Manuf.* **2020**, *32*, 100966. [CrossRef]
19. Bierwisch, C. Consistent Thermo-Capillarity and Thermal Boundary Conditions for Single-Phase Smoothed Particle Hydrodynamics. *Materials* **2021**, *14*, 4530. [CrossRef] [PubMed]
20. Minakov, A.; Morikawa, J.; Zhuravlev, E.; Ryu, M.; van Herwaarden, A.W.; Schick, C. High-speed dynamics of temperature distribution in ultrafast (up to 10^8 K/s) chip-nanocalorimeters, measured by infrared thermography of high resolution. *J. Appl. Phys.* **2019**, *125*, 054501. [CrossRef]
21. Hejmady, P.; van Breemen, L.C.A.; Hermida-Merino, D.; Anderson, P.D.; Cardinaels, R. Laser sintering of PA12 particles studied by in-situ optical, thermal and X-ray characterization. *Addit. Manuf.* **2022**, *52*, 102624. [CrossRef]

materials

MDPI

Article

Nanoparticle Tracing during Laser Powder Bed Fusion of Oxide Dispersion Strengthened Steels

Yangyiwei Yang [1], Carlos Doñate-Buendía [2], Timileyin David Oyedeji [1] and Bilal Gökce [2] and Bai-Xiang Xu [1,*]

1 Mechanics of Functional Materials Division, Institute of Materials Science, The Technical University of Darmstadt, D-64287 Darmstadt, Germany; yangyiwei.yang@mfm.tu-darmstadt.de (Y.Y.); timileyin.oyedeji@tu-darmstadt.de (T.D.O.)

2 Materials Science and Additive Manufacturing, School of Mechanical Engineering and Safety Engineering, University of Wuppertal, D-42119 Wuppertal, Germany; carlos.donate-buendia@uni-wuppertal.de (C.D.-B.); goekce@uni-wuppertal.de (B.G.)

* Correspondence: xu@mfm.tu-darmstadt.de

Abstract: The control of nanoparticle agglomeration during the fabrication of oxide dispersion strengthened steels is a key factor in maximizing their mechanical and high temperature reinforcement properties. However, the characterization of the nanoparticle evolution during processing represents a challenge due to the lack of experimental methodologies that allow in situ evaluation during laser powder bed fusion (LPBF) of nanoparticle-additivated steel powders. To address this problem, a simulation scheme is proposed to trace the drift and the interactions of the nanoparticles in the melt pool by joint heat-melt-microstructure–coupled phase-field simulation with nanoparticle kinematics. Van der Waals attraction and electrostatic repulsion with screened-Coulomb potential are explicitly employed to model the interactions with assumptions made based on reported experimental evidence. Numerical simulations have been conducted for LPBF of oxide nanoparticle-additivated PM2000 powder considering various factors, including the nanoparticle composition and size distribution. The obtained results provide a statistical and graphical demonstration of the temporal and spatial variations of the traced nanoparticles, showing $\sim$55% of the nanoparticles within the generated grains, and a smaller fraction of $\sim$30% in the pores, $\sim$13% on the surface, and $\sim$2% on the grain boundaries. To prove the methodology and compare it with experimental observations, the simulations are performed for LPBF of a 0.005 wt % yttrium oxide nanoparticle-additivated PM2000 powder and the final degree of nanoparticle agglomeration and distribution are analyzed with respect to a series of geometric and material parameters.

Keywords: additive manufacturing; laser powder bed fusion; selective laser melting; oxide dispersion strengthened steel; phase-field model; finite element simulation; nanoparticle interaction

Citation: Yang, Y.; Doñate-Buendía, C.; Oyedeji, T.D.; Gökce, B.; Xu, B.-X. Nanoparticle Tracing during Laser Powder Bed Fusion of Oxide Dispersion Strengthened Steels. *Materials* **2021**, *14*, 3463. https://doi.org/10.3390/ma14133463

Academic Editor: Aniello Riccio

Received: 10 May 2021
Accepted: 14 June 2021
Published: 22 June 2021

1. Introduction

Powder-based laser additive manufacturing techniques such as Laser Powder Bed Fusion (LPBF) [1] or Direct Energy Deposition (DED) [2] have been recently established as methods that allow the strengthening of metal alloys by modification of the microstructure [3]. Often, the strengthening is achieved by introducing lattice-matched nanoparticles within the surrounding matrix [1]. Another methodology used for strengthening is the introduction of exogenic dispersoids into the metallic matrix leading to the retardation of dislocation movements. Dislocations interact with the impenetrable dispersoids by the formation of a dislocation loop in between neighboring dispersoids only allowing dislocation loops with equilibrium diameter below the dispersoids interspacing to bypass the obstacle. This effect, known as the Orowan mechanism, describes that a fine dispersion results in efficient hardening [4]. Oxide dispersion strengthened (ODS) steels make use of this mechanism. The improved mechanical properties at high temperatures of ODS alloys can be related to the nature of the steel matrix as well as the composition, size, and distribution of the dispersoids (i.e., nanoparticles). Introduced nanoparticles in ODS steels are typically composed of yttrium-based oxides, exhibiting low solubility in the steel matrix and having a low potential for coarsening by Ostwald ripening [5]. In combination with the refining agent Titanium, nanoparticles with the chemical composition $Y_2Ti_2O_7$ [6], Y_2TiO_5 [6] are formed [7] with typical diameters of 2–3 nm in the 14YWT alloy [8]. The addition of aluminum enables the formation of various Y-Al-O-based compounds ($YAlO_3$, $Y_3Al_5O_{12}$) [9] leading to coarser dispersoids ($\sim$ 15 nm) and a reduced number density in the PM2000 alloy [8], resulting in increased ductility at the cost of lower strength.

The main fabrication route for ODS steels is the powder metallurgy route [10]. However, this expensive and complex fabrication route often leads to a low fracture toughness in the fabricated part and is inflexible in the part design [11]. As an alternative, additive manufacturing techniques are capable of producing ODS materials offering high solidification rates in combination with strong Marangoni forces within the melt pool, potentially leading to a homogeneous distribution of the introduced nanoparticles [12]. The distribution of these nanoparticles is highly influenced by the additive manufacturing technique employed and the parameters selected for processing due to differences in the melt pool dynamics. However, recent studies on this topic show that it is difficult to achieve the optimum nanoparticle size according to the Orowan mechanism due to segregation or agglomeration of the nanoparticles, which in turn was found to deteriorate the mechanical properties of the part [13]. Approaches to optimize the process such as evaluating the influence of the powder characteristics [14] and process parameters [15], or alternative additivation routes such as light mixing [16], improve the dispersion of the nanoparticles; however, they still lack control over the nanoparticle size.

The size and dispersion of the nanoparticles in the ODS steels are influenced during the steps that the initial powder undergoes until the fabrication of the ODS steel. First, the nanoparticles are supported on the steel microparticles. Ball milling of the yttrium-based nanoparticles with the steel powder is the most common approach to achieve it [17,18]. While widely employed, the control of the final nanoparticle size and degree of dispersion by this methodology is limited, and the steel microparticles size and shape can also be affected [19]. To address this drawback, alternative supporting procedures have been proposed [20] such as resonant acoustic mixing [21], solid–liquid reaction [22], or colloidal dielectrophoretic deposition [23,24]. Once the nanoparticle-additivated powder is obtained, the processing technique employed to generate the ODS steel samples and the experimental parameters selected determine the evolution of the nanoparticles, their final size, and dispersion in the ODS steel [12,25]. Even techniques with a similar working principle to LPBF like direct energy deposition (DED)—both laser additive manufacturing techniques—lead to differences in the nanoparticle agglomeration during processing due to the higher cooling rates of the melt pool achieved in LPBF that favor the preservation of the nanoparticle dispersion [12]. Even though these experimental observations provide an insight into the nanoparticle behaviors, further investigations would be required to completely understand the undergoing nanoparticle capture (nanoparticles trapped interior the microstructure), enrichment (local concentration increase of nanoparticles), and agglomeration processes [26–29]. Since the nanoparticles are not accessible during processing to perform in situ measurements, the combination of simulations with the experimental characterization of the nanoparticle-additivated powder and the generated ODS steel [13,30] represents the best approach to understand and control the LPBF processing of ODS steels.

Intuitively, the simulation of the LPBF processing of ODS steels and the underlying behaviors of nanoparticles, including their drift in the melt pool, captured during re-solidification, enrichment, and agglomeration, requires the proper modeling of the physical phenomena during the LPBF process and interactions among nanoparticles. It already remains a great challenge to model the underlying phenomena due to their sophisticated and interactive nature, covering a broad range of time and length scales. Notably, the thermal/mass transfers and material transformations (including melting, solidification and evaporation) dominate on a length scale of hundreds to thousands micrometers over a few dozen milliseconds. However, nanoparticle attraction and repulsion take place in a different spatial and temporal scale, around several tenths of nanometers and within microseconds. Finally, the manufacturing of an ODS steel sample is explained in a completely different scale of centimeters and takes hours or even days [31]. Resultant morphologies also reveal themselves in a multiscale fashion [32]. In this sense, the existent simulation schemes for the LPBF process are more or less subjected to strong simplifications and segregated modeling schemes, considering either only selected aspects or with the thermal history taken as input from separate numerical approaches. Those schemes usually feature a computational fluid dynamics method to simulate the thermal-fluid coupled spatial-transient evolution, including the arbitrary Lagrangian-Eulerian [33–36] and Lattice Boltzmann [37–39], incorporated with another method to simulate the accompanied microstructure evolution (mostly the polycrystalline re-solidification), such as the cellular automata [40–44] and phase-field [45]. Recently, we proposed a new phase-field model considering coupled processes among heat transfer, melt flow dynamics, and microstructure evolution (noted as "heat-melt-microstructure–coupled processes"), which shows the possibility to simulate the LPBF process in a unified and thermodynamic consistent route and the ability to recapitulate various experimental observations via simulation, such as high-gradient temperature field, tilted columnar grains, and lack-of-fusion pores with irregular shape [46].

On the other hand, very few works have been conducted regarding nanoparticle drift and interactions in the melt pool, which is, however, the central aspect to investigate the behaviors such as enrichment and agglomeration. Xu et al. [29,47] aimed to convey fundamental understandings regarding nanoparticle capture during re-solidification in order to obtain a nanocomposite-dispersed metal bulk, bringing the consideration of the Van der Waals (abbreviated as VDW hereinafter) effect, the Brownian effect, and thermodynamic analysis into the modeling of interactions in the nanocomposite-melt dispersion system. There are also experimental works revealing the VDW [48] and electrostatic interactions [49–51] among inorganic nanocomposites in the liquid metal. Nevertheless, those works disregard the influence of the driving effect from the melt flow dynamics when describing the nanocomposite-liquid/molten metal dispersion as well as the inter-particle interactions. In this sense, a simulation scheme combining the coupled phenomena (especially the melt pool dynamic and microstructure evolution) and the nanoparticle behaviors is still to be developed for the investigation of LPBF processing of ODS steels.

Joining our heat-melt-microstructure (HMM) coupled phase-field model with nanoparticle kinematics, we present and apply in this work a simulation scheme for tracing the nanoparticle drift and interactions in the melt pool during the LPBF process of the ODS steels. This proposed scheme is aimed to demonstrate the chronological and spatial variation of multiple traced nanoparticles with respect to various factors (such as chemical composition and size distribution), which enables graphical and statistical analysis on the nanoparticle migration and further effects, such as nanoparticle capture, enrichment, and agglomeration. The nanoparticle compositions used in the simulation are chosen based on our experimental studies on the additive manufacturing of ODS steels, including the use of Y_2O_3 [12,23] and YIG (yttrium iron garnet, $Y_3Fe_5O_{12}$) [23] nanoparticles. Influences from different types of nanoparticle size distribution, including the monomodal, normal, and log-normal distributions fitted from the experimental measurement [23], are also investigated and discussed in this work.

2. Models

2.1. Non-Isothermal Phase-Field Model for Stable LPBF Processes

To properly simulate the microstructure of a polycrystalline material, manufactured under a stable LPBF process, a conserved order parameter (OP) ρ is employed to represent the substance and atmosphere/pores; a set of non-conserved OPs ϕ_S and ϕ_L is employed to represent the solid and liquid phases, respectively; and a series of non-conserved OPs $\{\eta_\alpha\}$

are employed to represent the orientation distribution among the powders/grains. Two constraints should be applied on those OPs to properly recapture the reality (Figure 1a), i.e., substance constraint $(1-\rho)+\phi_\mathrm{L}+\phi_\mathrm{S}=1$ to restrict the existence of liquid as well as solid within the substance only, and the polycrystal constraint $(1-\phi_\mathrm{S})+\sum_\alpha \eta_\alpha=1$ to restrict polycrystalline orientations inside the solid substance. Notice that the substance constraint always satisfies when $\phi_\mathrm{L}=\rho-\phi_\mathrm{S}$. Therefore, only one ϕ to represent the solid substance and $(\rho-\phi)$ to represent the liquid substance is sufficient. The other constraint can be fulfilled via some practical numerical method, e.g., the Lagrange multiplier or the penalty method.

We consider a stable LPBF process (Figure 1a), i.e., without significant vaporization and resultant phenomena, such as scattering and keyholing. In this sense, the heat/mass transfer through various diffusion paths (e.g., surface, grain interior, and grain boundary) and melt flow, the dynamics of melt flow as well as the bubbles, the melting-solidification of the grains, and some inter-coupling effects, such as thermocapillary (Marangoni effect) and thermophoresis (Soret effect), would take the major role determining the resultant microstructure of the manufactured sample. The nonlinear kinetic system in simulating the stable LPBF processes is then adapted from our HMM-coupled phase-field model [46]. It is worth noting that the HMM-coupled phase-field model, derived under the thermodynamic-consistent framework, can be regarded as the combination of Navier-Stokes-Cahn-Hilliard (NSCH) and Navier-Stokes-Allen-Cahn (NSAC) systems with inter-coupling effects integrated. In this work, however, mentioned inter-coupling effects are tentatively dropped under the consideration of the computational stability, consumption, and complicity control in variants. These effects would be explicitly covered in our separate (e.g., the thermophoresis in [52]) or upcoming works. The adapted nonlinear kinetic system for the velocity field of the melt $\mathbf{u}$, the temperature field T, and fields of OPs ρ, ϕ and $\{\eta_\alpha\}$, eventually presents as

$$\begin{cases} \nabla\cdot\mathbf{u}=0, \\ \dfrac{\mathrm{D}\mathbf{u}}{\mathrm{D}t}=-\nabla p+\dfrac{1}{\mathrm{Re}}\nabla^2\mathbf{u}+\dfrac{1}{\mathrm{Fr}^2}\hat{\mathbf{g}}, \\ \dfrac{\mathrm{D}T}{\mathrm{D}t}+\dfrac{1}{\mathrm{Ste}_\phi}\dfrac{\mathrm{D}\phi}{\mathrm{D}t}=\nabla\cdot\left(\dfrac{1}{\mathrm{Pe}_T}\nabla T\right)+q_\mathbf{v}, \\ \dfrac{\mathrm{D}\rho}{\mathrm{D}t}=\nabla\cdot\left(\dfrac{1}{\mathrm{Pe}_\rho}\nabla\dfrac{\delta\mathscr{F}}{\delta\rho}\right), \\ \dfrac{\mathrm{D}\phi}{\mathrm{D}t}=-\dfrac{1}{\mathrm{Ac}_\phi}\dfrac{\delta\mathscr{F}}{\delta\phi}, \\ \dfrac{\mathrm{D}\eta_\alpha}{\mathrm{D}t}=-\dfrac{1}{\mathrm{Ac}_\eta}\dfrac{\delta\mathscr{F}}{\delta\eta_\alpha}, \end{cases} \tag{1}$$

with dimensionless quantities, namely the Reynolds number (Re), the Froude number (Fr), the Stefan number (Ste), the Péclet numbers for thermal (Pe_T) and mass (Pe_ρ) transfer, and Allen-Cahn numbers for melting-solidification (Ac_ϕ) and grain growth (Ac_η). Here, $\hat{\mathbf{g}}$ stands for the unit vector of gravitation direction. These quantities are employed not only to parameterize the nonlinear system but also to characterize the ratio between their corresponding physical processes to the chosen rate (by default, the characteristic rate of the fluid). Detailed parameterization of these quantities will be explained in Section 3.3.

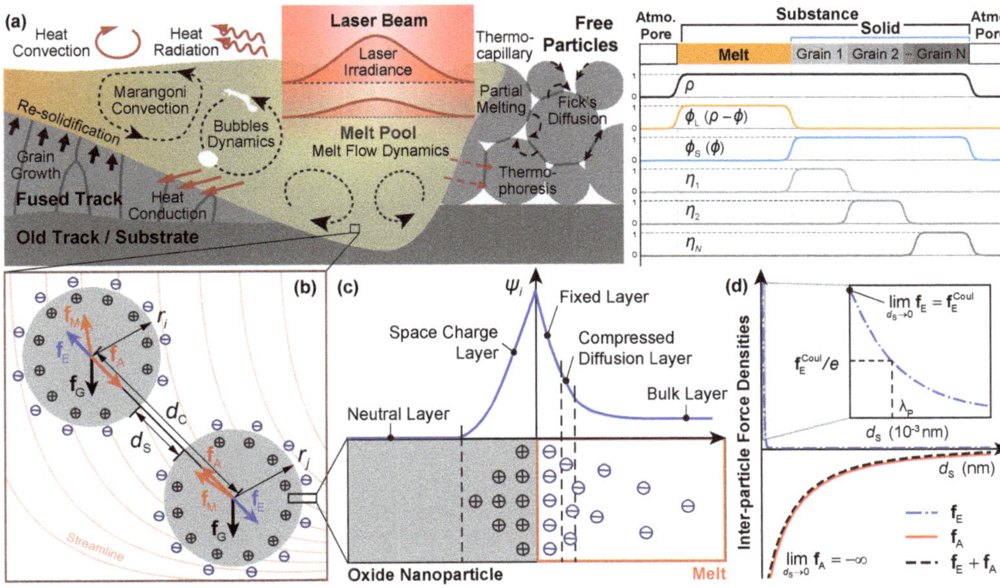

Figure 1. (**a**) Schematic of interactive physical phenomena during a stable LPBF process (left) and the order parameter profiles of the phase-field model (right); (**b**) force analysis of two oxide nanoparticles (labeled as *i* and *j*) in the melt pool with streamline and charges denoted; (**c**) schematic of charge distribution and potential profile across the oxide-melt interface based on explanations in [49]; (**d**) variation of the inter-particle force densities to the surface distance.

The explicit formulation of non-isothermal free energy $\mathscr{F}$ is formulated as

$$\begin{aligned}
\mathscr{F} &= \int_\Omega \Big[f_{\mathrm{ht}}(T,\rho,\phi,\{\eta_\alpha\}) + f_{\mathrm{loc}}(T,\rho,\phi,\{\eta_\alpha\}) + f_{\mathrm{grad}}(\nabla\rho,\nabla\phi,\{\nabla\eta_\alpha\}) \Big] \mathrm{d}\Omega. \\
f_{\mathrm{ht}}(T,\rho,\phi,\{\eta_\alpha\}) &= \xi\Big(\underline{A}\rho + \underline{G}\phi + \underline{B}\sum_\alpha \eta_\alpha\Big)\bigg\{ c_{\mathrm{r}}\Big[T\ln\frac{T}{T_{\mathrm{M}}} - (T - T_{\mathrm{M}})\Big] - \\
&\quad \Phi_{\mathrm{L}}\Big[\frac{T - T_{\mathrm{M}}}{T_{\mathrm{M}}}\mathscr{L}\Big]\bigg\}, \\
f_{\mathrm{loc}}(T,\rho,\phi,\{\eta_\alpha\}) &= w_{\mathrm{ss}}(T,\rho) + w_{\mathrm{sl}}(T,\rho,\phi) + w_{\mathrm{gr}}(T,\phi,\{\eta_\alpha\}), \\
f_{\mathrm{grad}}(T,\nabla\rho,\phi,\{\nabla\eta_\alpha\}) &= \frac{1}{2}T\kappa_\rho|\nabla\rho|^2 + \frac{1}{2}T\kappa_\phi|\nabla\phi|^2 \\
&\quad + \frac{1}{2}T\kappa_\phi|\nabla\rho - \nabla\phi|^2 + \frac{1}{2}T\kappa_\eta\sum_\alpha|\nabla\eta_\alpha|^2,
\end{aligned} \tag{2}$$

with

$$\begin{aligned}
w_{\mathrm{ss}}(T,\rho) &= \underline{C}(T)\Big[\rho^2(1-\rho)^2\Big], \\
w_{\mathrm{sl}}(T,\rho,\phi) &= \underline{H}(T)\Big[\rho^2 + 6(1-\rho)\{\phi^2 + (\rho-\phi)^2\} - 4(2-\rho)\{\phi^3 + (\rho-\phi)^3\} \\
&\quad + 3\{\phi^2 + (\rho-\phi)^2\}^2\Big], \\
w_{\mathrm{gr}}(T,\phi,\{\eta_\alpha\}) &= \underline{D}(T)\left[\phi^2 + 6(1-\phi)\sum_\alpha\eta_\alpha^2 - 4(2-\phi)\sum_\alpha\eta_\alpha^3 + 3\Big(\sum_\alpha\eta_\alpha^2\Big)^2\right],
\end{aligned}$$

and

$$\underline{C}(T) = \underline{C}_{\rm pt} - \underline{C}_{\rm cf}(T - T_{\rm M}),$$
$$\underline{D}(T) = \underline{D}_{\rm pt} - \underline{D}_{\rm cf}(T - T_{\rm M}),,$$
$$\underline{H}(T) = \underline{H}_{\rm pt} - \underline{H}_{\rm cf}(T - T_{\rm M})$$

where $T_{\rm M}$ is the melting temperature, $\mathscr{L}$ is the latent heat of the material, and $c_{\rm r}$ is the relative volumetric specific heat. Therefore, the variational derivatives of the free energy appearing in Equation (2) yield

$$\frac{\delta \mathscr{F}}{\delta \rho} = \frac{\partial f_{\rm ht}}{\partial \rho} + \frac{\partial f_{\rm loc}}{\partial \rho} - T\kappa_\rho \nabla^2 \rho - T\kappa_\phi(\nabla^2\rho - \nabla^2\phi),$$
$$\frac{\delta \mathscr{F}}{\delta \phi} = \frac{\partial f_{\rm ht}}{\partial \phi} + \frac{\partial f_{\rm loc}}{\partial \phi} - T\kappa_\phi \nabla^2 \phi + T\kappa_\phi(\nabla^2\rho - \nabla^2\phi),$$
$$\frac{\delta \mathscr{F}}{\delta \eta_\alpha} = \frac{\partial f_{\rm ht}}{\partial \eta_\alpha} + \frac{\partial f_{\rm loc}}{\partial \eta_\alpha} - T\kappa_\eta \nabla^2 \eta_\alpha.$$

Model parameters $\underline{A}, \underline{B}, \underline{G}, \underline{C}_{\rm pt}, \underline{C}_{\rm cf}, \underline{D}_{\rm pt}, \underline{D}_{\rm cf}, \underline{H}_{\rm pt}, \underline{H}_{\rm cf}$, as well as the gradient constants (κ_ρ, κ_ϕ, and κ_η), are obtained from given diffusive interface width and the experimentally measured interface energies. Coefficient ξ is employed to favor the determination of model parameters by fitting the experimental results (sufficiently summarized in Ref. [46] and supplementary information of Ref. [53]). $\Phi_{\rm L}$ is the interpolation function indicating the spatial landscape of the liquid/melt (see Section 3.3).

Finally, the thermal effect is equivalently treated as an internal heat source term $q_{\bf v}$ moving with the scan velocity $\mathbf{v}_{\rm l}$

$$q_{\bf v} = \Phi_{\rm ss}\beta P_{\rm l}\left\{\frac{\Pi}{\pi R_{\rm l}^2}\exp\left[-\Pi\frac{|\mathbf{x} - \mathbf{v}_{\rm l}t|^2}{R_{\rm l}^2}\right]\right\}\frac{2}{\zeta\sqrt{\pi}}\left(-\frac{|z - z_{\bf v}|^2}{\zeta^2}\right), \tag{3}$$

in which $P_{\rm l}$ is the laser beam power reaching the surface of the powder bed, β is the attenuation coefficient. $\Phi_{\rm ss}$ is the interpolation function for the substance. $\mathbf{x}$ is an arbitrary point on the projected plane of the laser beam on the powder bed surface, while z is an arbitrary depth from the plane. $(\mathbf{v}_{\rm l}t, z_{\bf v})$ is the moving center of the beam following the morphology of the surface. ζ is the characteristic penetration depth and normally takes the value of the powder bed thickness. Notice that parameter Π is utilized to adjust the concentration of the deposited power inside the circular beam spot with nominal radius $R_{\rm l}$, e.g., as suggested by the ISO standard [54], $\Pi = 2$, indicating 86.5% of the concentrated power within the spot. $P_{\rm l}$, $R_{\rm l}$, and the mode of the scan velocity $v_{\rm l} = |\mathbf{v}_{\rm l}|$ (scan speed) are thereby regarded as the major processing parameters of the laser scan in this work.

2.2. Nanoparticle Kinematics

The coupled evolution among polycrystalline microstructure, melt flow dynamics, and temperature transfer is calculated on the mesoscale (0.1–100 μm) using the phase-field model presented above. Since the size of nanoparticles is smaller by several orders, the possible influence of nanoparticles on these mesoscale effects is ignored. As an important part of nanoparticle kinematics, the drift effect of melt flow is inherited from the phase-field simulations. Moreover, nanoparticles can interact with each other by different mechanisms. Based on former research for dispersed non-metallic particles [29,47,55–57] neglecting the rotation, the kinematic equation for a dispersed rigid nanoparticle (labeled as i) with a volume V_i, the density ϱ_i, and the translation velocity $\mathbf{v}_i$ in the melt,

$$\varrho_i\frac{\mathrm{d}\mathbf{v}_i}{\mathrm{d}t} = \varrho_i\mathbf{g} + \mathbf{f}_{\rm M} + \mathbf{f}_{\rm A} + \mathbf{f}_{\rm E}, \tag{4}$$

in which the r.h.s. terms are the driving force densities due to the gravity, the melt flow, the VDW, and the electrostatic interaction, respectively (Figure 1b).

Considering the boundary of the nanoparticle as ∂V_i, the force density due to melt-flow driven is formulated as $\mathbf{f}_\mathrm{M} = \frac{1}{V_i}\int_{\partial V_i} \sigma \cdot \mathbf{n}\,\mathrm{d}S$, in which the Cauchy stress tensor of the fluid reads $\sigma = -p\mathbf{I} + \frac{1}{\mathrm{Re}}\nabla^2\mathbf{u}$. This is widely employed in simulating nanoparticle dispersion, including lattice-Boltzmann [57,58] and finite element method using static [59] or fictitious domain [60,61]. Assuming an infinitesimal size of the nanoparticle compared to the characteristic length scale of the melt, i.e., $V_i \to 0$, $\mathbf{f}_\mathrm{M}$ can be thus calculated as follows in this work according to the definition of the divergence

$$\begin{aligned} \mathbf{f}_\mathrm{M} &= \lim_{V_i \to 0} \frac{1}{V_i}\int_{\partial V_i} \sigma \cdot \mathbf{n}\,\mathrm{d}S = \nabla \cdot \sigma|_{\mathbf{x}_i} \\ &= \varrho_\mathrm{m}\left(\left.\frac{\mathrm{D}\mathbf{u}}{\mathrm{D}t}\right|_{\mathbf{x}_i} - \frac{1}{\mathrm{Fr}^2}\hat{\mathbf{g}}\right), \end{aligned} \tag{5}$$

where $\mathbf{x}_i$ is the center point of the nanoparticle, and $\mathrm{D}\mathbf{u}/\mathrm{D}t|_{\mathbf{x}_i}$ is the undisturbed time difference of the melt velocity at point $\mathbf{x}_i$.

The VDW and electrostatic force densities, depending on respective inter-particle potential U_A and U_E, can be formulated as

$$\mathbf{f}_\mathrm{A} = -\frac{1}{V_i}\nabla_d U_\mathrm{A}(d_\mathrm{C}),\ \mathbf{f}_\mathrm{E} = -\frac{1}{V_i}\nabla_d U_\mathrm{E}(d_\mathrm{C}), \tag{6}$$

where d_C is the inter-particle center distance, $d_\mathrm{C} = |\mathbf{x}_i - \mathbf{x}_j|$, distinguishing for the surface distance $d_\mathrm{S} = |\mathbf{x}_i - \mathbf{x}_j| - (r_i + r_j)$, as shown in Figure 1b. ∇_d represents the gradient operator in the inter-particle direction. Derived by Hamaker [62], the inter-particle potential of VDW attraction between nanoparticles i and j with corresponding radius r_i and r_j is

$$U_\mathrm{A}(d_\mathrm{C}) = -\frac{A_\mathrm{p:m}}{6}\left\{\frac{2r_ir_j}{d_\mathrm{C}^2 - (r_i + r_j)^2} + \frac{2r_ir_j}{d_\mathrm{C}^2 - (r_i - r_j)^2} + \ln\left[\frac{d_\mathrm{C}^2 - (r_i + r_j)^2}{d_\mathrm{C}^2 - (r_i - r_j)^2}\right]\right\}, \tag{7}$$

where $A_\mathrm{p:m}$ is referred to as the Hamaker coefficient of the nanoparticle in a medium, which is dependent on the permittivity of involved materials and the intervening medium. According to Dzyaloshinskii-Lifshitz-Pitaevskii interpretation [63,64], in which nanoparticles are treated as continuous media (rather than atomic structure as [62]), and inter-particle forces are derived in terms of permittivities and refractive indices, the non-retarded (instantaneous) Hamaker coefficient for VDW interactions between two nanoparticles of the same material through a medium (denoted as $A_\mathrm{p:m}$) is estimated as

$$A_\mathrm{p:m} = \frac{3}{4}k_\mathrm{B}T\left(\frac{\varepsilon_\mathrm{p} - \varepsilon_\mathrm{m}}{\varepsilon_\mathrm{p} + \varepsilon_\mathrm{m}}\right)^2 + \frac{3\hbar\omega_\mathrm{e}}{16\sqrt{2}}\frac{\left(n_\mathrm{p}^2 - n_\mathrm{m}^2\right)^2}{\left(n_\mathrm{p}^2 + n_\mathrm{m}^2\right)^{3/2}} \tag{8a}$$

with the permittivities ε_p and ε_m, and the refractive indices n_p and n_m of the nanoparticle and the medium, respectively. k_B and $\hbar$ are Boltzmann constant and reduced Planck constant, while $\omega_\mathbf{e}$ is the electronic absorption frequency, ranging in $3 \sim 5 \times 10^{15}$ Hz. It is worth noting that Equation (8a) is not employable due to the difficulties in the practical measurement of permittivity and refractive index of the molten metal. As an alternative, $A_\mathrm{p:m}$ is calculated from the ones for nanoparticle ($A_\mathrm{m:v}$) and medium ($A_\mathrm{p:v}$) that obtained separately in the vacuum, according to the following combining relation [48,65]

$$A_\mathrm{p:m} = \left(\sqrt{A_\mathrm{p:v}} - \sqrt{A_\mathrm{m:v}}\right)^2, \tag{8b}$$

where $A_{\text{p:v}}$ can be obtained from experimentally measured optical data [56], while $A_{\text{m:v}}$ for dispersed metallic particle/droplet in the vacuum is derived from Dzyaloshinskii-Lifshitz-Pitaevskii theory by taking the estimated frequency-dependent permittivity for metals as $\varepsilon_{\text{m}}(\omega) \approx 1 - \omega_{\text{e}}^2/\omega^2$ with two critical conditions, i.e., $\varepsilon_{\text{m}}(0) = \infty$ and $\varepsilon_{\text{m}}(\infty) = 1$, which is valid for plasma and metal. Then, $A_{\text{m:v}}$ eventually yields

$$A_{\text{m:v}} = \frac{3}{4}k_{\text{B}}T + \frac{3\hbar\omega_{\text{e}}}{16\sqrt{2}}. \tag{8c}$$

Obviously, $A_{\text{p:m}}$ obtained from either Equations (8a) and (8b) is always positive with a typical value ranging in $10^{-19} \sim 10^{-20}$ J, demonstrating an ever attractive VDW forces between two nanoparticles of the same material through a melt (taking negative value as attraction and positive one as repulsion for all nanoparticles if without further interpretation).

Unlike in the aqueous solution, electrostatic repulsion in the molten metal has rarely been studied. Although there are works [29,47] assuming negligible electrostatic interaction due to the strong screening effect of background free electrons in the melt, there is reported evidence of charged surface on the oxide nanoparticle in the liquid metal [49,50]. Due to accompanied electron defects (like quasi-free electrons or holes) in the oxide, electron flow occurs across the interface when the oxide nanoparticles are in contact with the liquid metal, resulting in the surface potential ψ_i and corresponding profiles at the interface [49,51], as shown in Figure 1c. Combining the above viewpoints from existing studies, the following assumptions are made in the work to model the electrostatic interaction between nanoparticles through melt:

1. Electrons in the molten metal behaves as a free electron gas with a density n_{e}, receiving only the contributions from valence electrons of all metallic elements in the melt.
2. Charged surface exists on the oxide nanoparticles in the melt, even though its effect is weak or absent due to the strong screening effect. The value of this surface charge ψ_i is assumed to be equal to that obtained from a neutral aqueous dispersion.

Then, the screened-Coulombic potential between two nanoparticles, applied for electrostatic interaction in the electron-screened system (like plasma), is adopted from [66], i.e.

$$U_{\text{E}}(d_{\text{C}}) = Z\frac{r_i r_j}{d_{\text{C}}}\exp\left[\frac{d_{\text{C}} - (r_i + r_j)}{\lambda_{\text{P}}}\right], \tag{9}$$

where the interaction coefficient Z is analogous to the Hamaker coefficient, reading as

$$Z = 4\pi\varepsilon_0\psi_i\psi_j, \tag{10}$$

which also presents a positive value for nanoparticles of the same material in the melt, demonstrating an ever repulsive electrostatic interaction. λ_{P} is the plasma Debye length, which depends on the free electron density n_{e}, i.e.,

$$\lambda_{\text{P}} = \sqrt{\frac{\varepsilon_0 k_{\text{B}} T_{\text{M}}}{n_{\text{e}} e_0^2}}, \quad n_{\text{e}} = \varrho_{\text{m}} N_{\text{A}} \frac{\sum_l c_l z_l}{\sum_l c_l m_l},$$

where z_l, c_l, and m_l are the valence electron number, the atom fraction and the atom mass of the element l, respectively. N_{A} is the Avogadro constant, ε_0 is the vacuum permittivity, and e_0 is the elementary charge. Taking the formulation of the interaction potentials in Equations (7) and (9), the VDW and electrostatic force densities are formulated as

$$\mathbf{f}_\text{A} = -\frac{32A_\text{p:m}d_\text{C}(r_ir_j)^3}{3V_i\left[d_\text{C}^4 - 2d_\text{C}^2\left(r_i^2 + r_j^2\right) + \left(r_i^2 - r_j^2\right)^2\right]^2}\frac{\mathbf{x}_i - \mathbf{x}_j}{d_\text{C}},$$
$$\mathbf{f}_\text{E} = Z\frac{r_ir_j(d_\text{C} + \lambda_\text{P})}{V_i\lambda_\text{P}d_\text{C}^2}\exp\left[-\frac{d_\text{C} - (r_i + r_j)}{\lambda_\text{P}}\right]\frac{\mathbf{x}_i - \mathbf{x}_j}{d_\text{C}}. \tag{11}$$

According to its physical meaning, λ_P characterizes the strength of the screening effect due to the electrostatic interaction received by one nanoparticle, which is reflected by the decaying length scale of the $\mathbf{f}_\text{E}$, as shown in Figure 1d. It is calculated that for the alloy such as PM2000, λ_P is at the scale of 10^{-3} nm, indicating a very short-range electrostatic repulsion. In addition, $\mathbf{f}_\text{E}$ trends to the ordinary Coulombic potential $\mathbf{f}_\text{E}^\text{Coul}$ with a finite value when two nanoparticles are getting in touch, while $\mathbf{f}_\text{A}$ tends to infinity. This demonstrates that the attractive VDW interaction dominates the resultant effect between two nanoparticles, which would destabilize the dispersion system from a perspective of colloidal science [29,48,65] and lead to agglomeration of the nanoparticles. This is against the preference of forming a homogeneous oxide dispersion during the LPBF process. Therefore, the validation of the proposed model relies on further experimental insights into the interactive behavior of the nanoparticle in the molten/liquid metal, which would be covered in our future works.

3. Methods

3.1. Numerical Scheme and Implementation

Assuming negligible counter-effect of the nanoparticles on the melt flow, the simulation is designed in a subsequent scheme: an HMM-coupled phase-field simulation for the LPBF process and a subsequent nanoparticle kinematics simulation with information received from the phase-field simulation, as shown in Figure 2. The information includes the initial condition (IC) of the powder bed as a set $\mathbb{V}(\mathbf{X}_I, R_I)$ with the center $\mathbf{X}_I$ and the radius R_I of the powder I to create IC (centers and radii) for the additivated nanoparticles, and nodal values (ρ^J, ϕ^J, $\{\eta^J\}$ and $\mathbf{f}_\text{M}^J$ of nodal J) of every time-step to provide the melt-flow driving force as well as the on-site phase information. Due to the infinitesimal-volume assumption, the drifted nanoparticles are represented and traced by corresponding center position $\mathbf{x}_i$. The trajectories of the nanoparticles are then calculated numerically by discretizing the kinematic model outlined in Section 2.2 in the backward differences fashion:

$$\mathbf{v}_i|_t = \mathbf{v}_i|_{t-\delta t} + \left[\mathbf{g} + \frac{1}{\varrho_i}(\mathbf{f}_\text{M} + \mathbf{f}_\text{A} + \mathbf{f}_\text{E})\right]\delta t$$
$$\mathbf{x}_i|_t = \mathbf{x}_i|_{t-\delta t} + \frac{1}{2}[\mathbf{v}_i|_{t-\delta t} + \mathbf{v}_i|_t]\delta t. \tag{12}$$

Note here the time difference δt should be by default no larger than the time interval Δt of the phase-field simulation to ensure the accuracy of the tracing. This tracing program repeats along with the phase-field simulations till reaching the stop criteria, i.e., the end time of simulation or interrupted due to non-converge situation during finite element method calculation. In addition to the trajectories, another important piece of information is the relative position of a nanoparticle in mesoscopic microstructure, for instance, grain interior, grain boundary region, pore, or surface. The position of a nanoparticle can change during its drift and is thus described by a chronological variable, which is termed here as "position indicator".

The phase-field model is numerically implemented via the finite element method within the program `NIsoS` developed by authors based on `MOOSE` framework (Idaho National Laboratory, ID, USA) [67]. Four-node quadrilateral Lagrangian elements are chosen to mesh the geometry. The Cahn-Hilliard equation is solved in a split way [68,69].

A transient solver with preconditioned Jacobian-Free Newton-Krylov (PJFNK) method and second-order backward Euler algorithm has been employed to solve the non-isothermal phase-field problems. Adaptive meshing and time-stepping schemes are used to reduce the computation costs. In addition, to stabilize the calculation of NSCH and NSAC systems, streamline-upwind Petrov-Galerkin and pressure-stabilized Petrov-Galerkin methods are introduced associated with the weak forms of the Navier-Stokes equations [70]. For now, the subsequent nanoparticle tracing program is coded by `Python` (ver. 3.7.10), which is independent of `NIsoS`. It is planned to integrate the tracing program as a subroutine of the phase-field simulation program in the future.

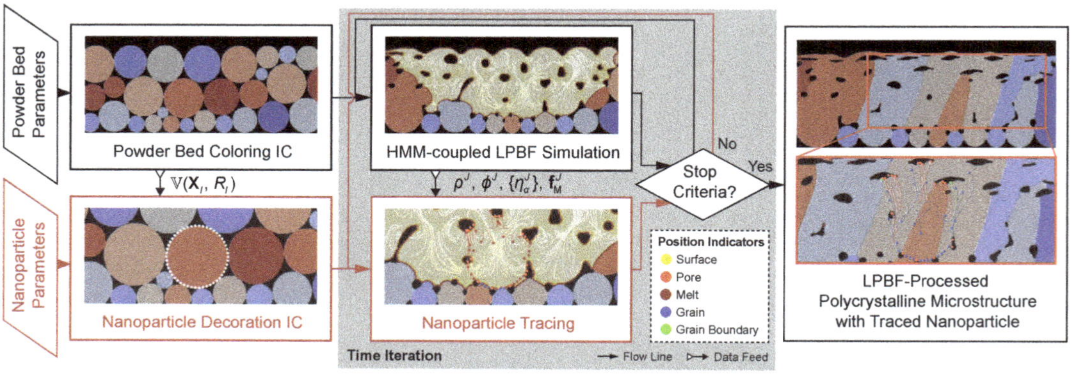

Figure 2. Workflow of the simulation scheme, including a phase-field simulation and a subsequent nanoparticle tracing program.

3.2. Simulation Setup

As a preliminary step, we apply first our nanoparticle tracking scheme for a 2D phase/field simulation of LPBF following our former work [46], which can recapture certain characteristic powder bed features, e.g., particles with multiple sizes and various pores due to the particle packing. The simulation domain has a size of 500×100 µm. Particles inside the domain are generated with the random close packing procedure. Due to the uncertainty of the initial grain structure of a single particle, we simply treat each particle as a monocrystal with a unique random orientation following the reported simulation works [46,71]. With the help of the minimum coloring algorithm and grain tracking algorithm [53,72], five η_α are sufficient to uniquely represent all the particles/grains for these simulations. The zero Neumann boundary condition (BC) for ρ and zero Dirichlet BC for $\mathbf{u}$, representing a close BC for the system, are applied on the boundary $\Gamma = \partial\Omega$ of the whole simulation domain Ω,

$$\nabla\rho|_\Gamma \cdot \hat{\mathbf{n}} = 0, \quad \mathbf{u}|_\Gamma = \mathbf{0}, \tag{13}$$

where $\hat{\mathbf{n}}$ is the normal vector of the boundary Γ, and $\mathbf{0}$ is the null vector. The heat convective BC allows heat dissipation as heat convection with the atmosphere on the top boundary Γ'

$$-k\nabla T|_{\Gamma'} \cdot \hat{\mathbf{n}} = h(T|_{\Gamma'} - T_\text{E}), \tag{14}$$

where h is the convective coefficient and T_E is the environment temperature. The Dirichlet BC on temperature with a fixed pre-heating temperature T_P is applied on the rest of the boundary Γ'' ($\Gamma = \Gamma' \cup \Gamma''$) to emulate the contact with a semi-infinite heat reservoir (e.g., the substrate)

$$T|_{\Gamma''} = T_\text{P}, \tag{15}$$

which helps to restrain the melt pool size for better demonstration in this work.

In addition, the pinning pressure BC on the top-left corner C ($C \in \Gamma'$) is applied on the hydrodynamic pressure p as

$$p|_C = 0 \tag{16}$$

to avoid the difficulties associated with the non-trivial nullspace of the operator pre-specified in the PJFNK solver as suggested in Ref. [70].

3.3. Parameters and Properties

The dimensionless quantities are employed in the nonlinear kinetic system for HMM-coupled phase-field simulations as explained in Section 2.1. Following the conventions of NSCH and NSAC system, they are formulated explicitly as

$$\begin{aligned} &\mathrm{Re} = \frac{\varrho\bar{v}\bar{l}}{\nu}, \quad \mathrm{Fr} = \sqrt{\frac{\varrho\bar{v}}{b\bar{t}}}, \quad \mathrm{Ste} = \frac{c_\mathrm{r}\bar{T}}{\mathscr{L}}, \\ &\mathrm{Pe}_\rho = \frac{\bar{v}\ell\bar{l}}{\bar{\gamma}M}, \quad \mathrm{Pe}_T = \frac{\bar{v}\bar{l}c_\mathrm{r}}{k}, \quad \mathrm{Ac}_\phi = \frac{\bar{v}\ell}{\bar{\gamma}\bar{l}L_\phi}, \quad \mathrm{Ac}_\eta = \frac{\bar{v}\ell}{\bar{\gamma}\bar{l}L_\eta}, \end{aligned} \tag{17a}$$

where $\bar{v}$, $\bar{l}$ and $\bar{T}$ are the characteristic velocity, length and temperature of the melt flow, while ℓ is characteristic width of the diffusive interface corresponding to the given characteristic surface tension $\bar{\gamma}$. ν is dynamic viscosity, k is the thermal conductivity, and M is the isotropic diffusivity. L_ϕ and L_η are the isotropic Allen-Cahn mobility of the solid-liquid interface and grain boundaries. b is the magnitude of the resultant body forces acting on the melt flow. For convenience, we use a set of simpler reference quantities to re-define those characteristic quantities by substituting the following relations in Equation (17a):

$$\bar{v} = \frac{\bar{l}}{\bar{t}}, \quad \bar{T} = T_\mathrm{M}, \quad \bar{\gamma} = \sqrt{T_\mathrm{M}\kappa_\rho^{T_\mathrm{M}}\underline{C}_\mathrm{pt}^{T_\mathrm{M}}}, \quad \ell = \sqrt{\frac{T_\mathrm{M}\kappa_\rho^{T_\mathrm{M}}}{\underline{C}_\mathrm{pt}^{T_\mathrm{M}}}}.$$

Notice that the $\underline{C}_\mathrm{pt}^{T_\mathrm{M}} = \kappa_\rho^{T_\mathrm{M}} T_\mathrm{M}/\bar{l}^2$ which is the model parameter obtained at the reference temperature T_M. $\kappa_\rho^{T_\mathrm{M}}$ is the gradient model parameter at a reference temperature T_M. Then, dimensionless quantities in Equation (17a) can be thereby modified as

$$\begin{aligned} &\mathrm{Re} = \frac{\varrho\bar{l}^2}{\bar{t}\nu}, \quad \mathrm{Fr} = \sqrt{\frac{\varrho\bar{l}}{b\bar{t}^2}}, \quad \mathrm{Ste} = \frac{c_\mathrm{r}T_\mathrm{M}}{\mathscr{L}}, \\ &\mathrm{Pe}_\rho = \frac{\bar{l}^2}{M\bar{t}\underline{C}_\mathrm{pt}^{T_\mathrm{M}}}, \quad \mathrm{Pe}_T = \frac{\bar{l}^2 c_\mathrm{r}}{\bar{t}k}, \quad \mathrm{Ac}_\phi = \frac{1}{L_\phi\bar{t}\underline{C}_\mathrm{pt}^{T_\mathrm{M}}}, \quad \mathrm{Ac}_\eta = \frac{1}{L_\eta\bar{t}\underline{C}_\mathrm{pt}^{T_\mathrm{M}}}. \end{aligned} \tag{17b}$$

Notice here that material properties c_r, k, ϱ, ν, should be phase-dependent and thereby formulated in a direct interpolated fashion as

$$\begin{aligned} c_\mathrm{r} &= \Phi_\mathrm{S}c_\mathrm{S}^p\varrho_\mathrm{S} + \Phi_\mathrm{L}c_\mathrm{L}^p\varrho_\mathrm{L} + \Phi_\mathrm{at}c_\mathrm{at}^p\varrho_\mathrm{at}, \\ k &= \Phi_\mathrm{S}k_\mathrm{S} + \Phi_\mathrm{L}k_\mathrm{L} + \Phi_\mathrm{at}k_\mathrm{at}, \\ \varrho &= \Phi_\mathrm{S}\varrho_\mathrm{S} + \Phi_\mathrm{L}\varrho_\mathrm{L} + \Phi_\mathrm{at}\varrho_\mathrm{at}, \\ \nu &= \Phi_\mathrm{S}\nu_\mathrm{S} + \Phi_\mathrm{L}\nu_\mathrm{L} + \Phi_\mathrm{at}\nu_\mathrm{at}, \end{aligned} \tag{18}$$

where $c_{(\cdot)}^p$, $k_{(\cdot)}$, $\varrho_{(\cdot)}$, and $\nu_{(\cdot)}$ are respectively the specific enthalpy, thermal conductivity, density, and dynamic viscosity of the corresponding phase. Similarly, the effective value of mobility L_η and M through possible paths are adopted correspondingly from the self-diffusivities $D^{\mathrm{eff}}_{(\cdot)}$ and grain boundary mobility $G^{\mathrm{eff}}_{\mathrm{gb}}$ [46,53,73,74], i.e.,

$$\begin{aligned} M &= \frac{1}{\partial^2 f_{\mathrm{loc}}/\partial\rho^2|_{\rho=1}}\Big(\Phi_{\mathrm{ss}}D_{\mathrm{ss}} + \Phi_{\mathrm{at}}D_{\mathrm{at}} + \Phi_{\mathrm{sf}}D_{\mathrm{sf}} + \Phi_{\mathrm{gb}}D_{\mathrm{gb}}\Big), \\ L_\eta &= \frac{G_{\mathrm{gb}}\gamma_{\mathrm{gb}}}{T\kappa_\eta}, \end{aligned} \tag{19}$$

while the quantity of L_ϕ is tentatively given as $20/\bar{t}C^{T_{\mathrm{M}}}_{\mathrm{pt}}$, which makes resultant $1/\mathrm{Ac}_\phi$ sufficiently larger than $1/\mathrm{Ac}_\eta$ to emulate a relatively faster melting-solidifying process than grain growth [46]. This is due to the lack of a quantitative description of migration mobility of the liquid. The subscript represents the quantities of the corresponding phases, e.g., 'ss' as the substance, 'at' as the atmosphere/pore, 'sf' as the surface, 'gb' as the grain boundary, 'S' as the solid and 'L' as the liquid/melt. Then, the interpolating functions Φ_{ss}, Φ_{at}, Φ_{gb}, Φ_{sf}, Φ_{S}, and Φ_{L} can be simply formulated as

$$\begin{aligned} &\Phi_{\mathrm{ss}} = \rho^3\Big(10 - 15\rho + 6\rho^2\Big), && \Phi_{\mathrm{at}} = 1 - \rho^3\Big(10 - 15\rho + 6\rho^2\Big), \\ &\Phi_{\mathrm{sf}} = 16\rho^2(1-\rho)^2, && \Phi_{\mathrm{gb}} = 16\sum_{i\neq j}\eta_i^2\eta_j^2, \\ &\Phi_{\mathrm{S}} = \phi^3\Big(10 - 15\phi + 6\phi^2\Big), && \Phi_{\mathrm{L}} = \Phi_{\mathrm{ss}} - \Big[\phi^3\Big(10 - 15\phi + 6\phi^2\Big)\Big]. \end{aligned} \tag{20}$$

Note that the constraints on the OPs should be also applied on the interpolation functions, i.e., $1 = \Phi_{\mathrm{ss}} + \Phi_{\mathrm{at}}$ and $\Phi_{\mathrm{ss}} = \Phi_{\mathrm{L}} + \Phi_{\mathrm{S}}$.

4. Results and Discussion

Simulation results are presented here for a single scan LPBF nanoparticle-decorated PM2000 alloy in an argon atmosphere. The composition of the PM2000 adopted in this work (presented in Table 1 following our former experimental investigations [12,23]) presents a ferritic structure in the high-temperature range (from 1000 K to the melting temperature) without solid-state phase transition, according to the Fe-Cr-Al ternary phase diagram reported in Ref. [75]. Therefore, the material properties for ferritic PM2000 alloy displayed in Table 2 are employed for the simulations. The reference length scale selected for the simulation is $\bar{l} = 1$ µm, and the time scale $\bar{t} = 1$ µs. The LPBF parameters are selected according to the LPBF experiments in [12]. The characteristic radius of the beam is set as $R_{\mathrm{l}} = 80$ µm, the penetration depth ζ is defined by the thickness of the powder bed, and the laser power and scanning speed are $P_{\mathrm{l}} = 160$ W and $v_{\mathrm{l}} = 800$ mm/s, respectively. The attenuation coefficient of the laser is $\beta = 0.65$. The environment temperature T_{E} as well as the pre-heating temperature T_{P} are both set to 353 K.

Table 1. Chemical composition of the PM2000 alloy as measured by XRF (only elements present in an amount ≥ 0.01 wt % are shown) [12,23].

	Fe	Cr	Al	Ti	Ni	Si	Cu
wt %	Bal.	20.40	3.94	0.58	0.10	0.03	0.01

The single scan LPBF of the monomodal Y_2O_3 (15 nm in radius) nanoparticle-additivated PM2000 powder is first discussed. Chronological microstructure evolution along with the traced nanoparticles and their trajectory are depicted in Figure 3. To provide a further insight into the steel and nanoparticle evolution during processing, three PM2000 particles from different locations with radii of 17–19 µm are chosen as the "parent particle" and marked as 'A', 'B', and 'C'. Fifty equispaced nanoparticles are placed on their surfaces in order to simulate a perfectly homogeneous initial nanoparticle dispersion. Parameters for nanoparticle interactions, including the Hamaker coefficient and the surface potential can be found in

Table 3. As the HMM-coupled phase-field simulations [46] show, the microstructure evolution of LPBF-processed PM2000 features multiple phenomena, including melt flow convection as well as accompanied behaviors of pores/bubbles, partially melted particles, re-solidification, and sintering of particles/grains, resulting in columnar grains and trapped irregular-shape pores due to lack of fusion. Those phenomena are also highlighted in Figure 3a_1–a_4. The focus here is on how the nanoparticles drift and eventually migrate under the influence of those effects. Initially, all the nanoparticles are located onto their corresponding parent particles before the laser interaction and generation of the melt pool (Figure 3a_1). Once fully immersed by the melt, the nanoparticle drift is firstly driven by the melt flow, i.e., there is a driving force along the tangent direction of the streamline (Figure 3a_2,b_1–d_1. This effect is illustrated in Figure 1b_1–b_4). Due to the sufficiently large spacing, other interaction contributions among nanoparticles are negligible. After the initial drift due to the melt front propagation, the evolution of the nanoparticles is associated with different physical phenomena depending on their location and surroundings. Nanoparticles around a pore/bubble, which are identified as the "bubble-carried" ones, would follow the floating, deforming, or even splitting of bubbles (Figures 3a_2,b_2–d_2,b_3–d_3). It is also possible that the melt flow brings multiple nanoparticles into a narrow region at the time, which is already within the range of nanoparticle interactions (specifically the VDW attraction since the electrostatic repulsion has an even shorter range around 10^{-3} nm). In this sense, the trajectory of certain nanoparticles would be redirected abruptly. Meanwhile, nanoparticles near the bottom of the melt pool present very little migration comparing to others. The reason can be the very short immersing time (thus less driven by melt-flow) or the partial melting of the corresponding parent particles. Next, when the laser front scans away, the local temperature drops and, once below T_M, the re-solidification occurs, forming the tail of the melt pool. Once the re-solidifying front goes through the migrating nanoparticles, they are immediately captured and become either interior or grain boundary (GB) nano-inclusions, while the ones carried by the bubbles stay as they are and become pore-trapped nanoparticles. Nanoparticles can also be found on the surface driven by either melt flow or emerging bubbles (Figure 3a_3). Figure 3d_2–d_4 present an additional case where nanoparticles, initially located on the surface, migrate in accordance with the deformation of the surface morphology. During this process, the nanoparticles with a relatively higher speed (driven by the melt flow) might enter the melt pool and become the in-grain nano-inclusions, while others remain on the surface. Finally, Figure 3a_4 presents traced nanoparticles with all sorts of position indicators (denoted by colors) and, notably, the locally enriched nano-inclusions and several ones with overlapped trajectories, implying potential agglomeration effects. One can readily tell from Figure 3a_2,a_3 that such enrichment majorly can be attributed to the melt-flow driving force, where multiple nanoparticles follow similar trajectories governed by the transient streamlines. However, it is worth noting that the overlap of the trace markers (indicating the center locations of the nanoparticles rather than the sizes) do not sufficiently reflect nanoparticle agglomeration effects, which should be explicitly determined by an adjacency test on the real scale of the nanoparticles' size. This will be discussed in the following demonstrations.

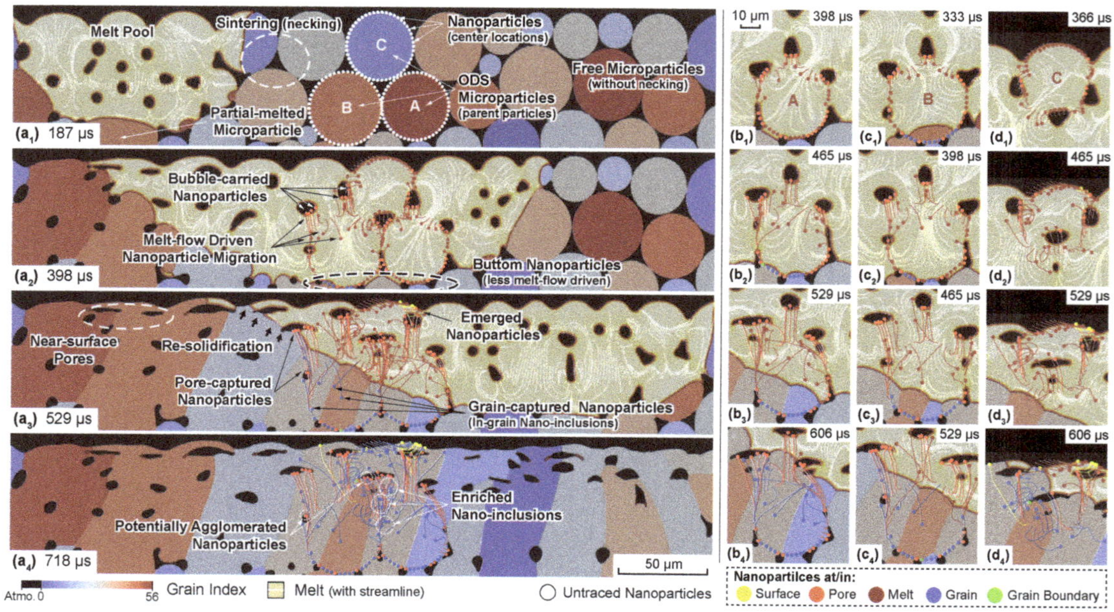

Figure 3. Simulation results of an LPBF-processed ferritic PM2000 powder bed with $P_l = 160$ W and $v_l = 800$ mm/s, in which three particles are marked and decorated with Y_2O_3 nanoparticles. Transient microstructure, nanoparticle tracing, and feature phenomena are presented at **(a$_1$)** 197 μs, **(a$_2$)** 398 μs, **(a$_3$)** 529 μs, **(a$_4$)** 718 μs . Trajectories of nanoparticles are respectively depicted for nanoparticles initially located on different parent particle at corresponding timestep, i.e., on parent particle A at **(b$_1$)** 398 μs, **(b$_2$)** 465 μs, **(b$_3$)** 529 μs, **(b$_4$)** 606 μs; on parent particle B at **(c$_1$)** 333 μs, **(c$_2$)** 398 μs, **(c$_3$)** 465 μs, **(c$_4$)** 529 μs; and on parent particle C at **(d$_1$)** 366 μs, **(d$_2$)** 465 μs, **(d$_3$)** 529 μs, **(d$_4$)** 606 μs. Notice that the trace markers at the end of the trajectory only indicate the central location of the nanoparticle at the time rather than the size.

Table 2. Material properties of the ferritic PM2000 alloy and Ar atmosphere, employed in the simulations.

Properties	Expressions (T in K)	Units	References
T_M	~ 1756.15	K	[76]
γ_{sf}^{exp}	$1.63 - 4.49 \times 10^{-3}(T - T_M)$ *	J/m^2	[77,78]
γ_{gb}^{exp}	$0.28 - 7.74 \times 10^{-3}(T - T_M)$ *	J/m^2	[77,78]
D_{sf}	$10 \exp(-2.41 \times 10^5/\Re T)$ †	m^2/s	[77]
D_{gb}	$1.1 \times 10^{-2} \exp(-1.74 \times 10^5/\Re T)$ †	m^2/s	[77]
D_{ss}	$1.8 \times 10^{-5} \exp(-2.08 \times 10^5/\Re T)$ †	m^2/s	[77]
G_{gb}	$5.36 \exp(-3.54 \times 10^5/\Re T)$	m^4/(J s)	[79]
k_{ss}	$30.841 + 0.011(T - T_M)$	J/(s m K)	[76]
k_{at}	~ 0.06	J/(s m K)	[80]
h_{at}	~ 100	J/(s m^2 K)	
c_{ss}^p	$908.596 + 0.323(T - T_M)$	J/(kg K)	[76]
c_{at}^p	520	J/(kg K)	[81]
$\mathscr{L}$	2.4×10^9 †	J/m^3	[82]
ϱ_{ss}	7180	kg / m^3	[76]
ϱ_{at}	1.38	kg / m^3	
ν_{ss}	$\sim 5.33 \times 10^{-3}$ ‡	(J s)/m^3	[83]
ν_{at}	$\sim 7.53 \times 10^{-5}$	(J s)/m^3	[84]

* Temperature-dependent data form [78] and scaled based on the value at T_M from [77]. † Data from ferrite. ‡ Linearly interpolated from temperature-dependent measurements on Fe-Cr melt with 21 at% Cr.

Table 3. Material properties and interaction coefficients of the Y_2O_3 and YIG nanoparticles.

	ϱ_i (kg/m^3)	$A_{p:v}$ (10^{-20} J)	$A_{p:m}$ (10^{-20} J)	ψ_i (mV)	Z (10^{-14}N)
Y_2O_3	5.01×10^3	14.0	5.43	26	7.52
YIG	5.17×10^3	24.2	1.33	59	38.73

Figure 4a,b presents the interactive forces vs. surface distance between two nanoparticles with different radii and composition, which are Y_2O_3 and YIG, their properties and related coefficients are listed in Table 3. Notice here the Hamaker coefficients in the melt for both Y_2O_3 and YIG are calculated according to Equation (8b) using the coefficients in vacuum for both compositions and the melt, where $A_{\mathrm{m:v}} = 37 \times 10^{-20}$ J is calculated by Equation (8c). It can be observed that, for the same surface distance, increasing nanoparticle radius leads to a higher VDW attraction as well as electrostatic repulsion. Comparing nanoparticles of the same size but different composition, YIG nanoparticles present less VDW attraction and more electrostatic repulsion compared to Y_2O_3 ones. Nevertheless, the interaction range of the electrostatic repulsion is limited to the 10^{-3} nm scale, and this interaction highly decreases in the nanoscale due to strong screening effects from background free electrons (reflected by very small λ_{P}), where the VDW attraction still takes the major role. This fact explains the instability of homogeneously dispersed nanoparticles on the 10 nm scale, as explained in Section 2.2. In the case of VDW attraction, it presents a considerable decay when $d_{\mathrm{S}} \geq 2$ nm, hence its influence over the inter-nanoparticle attractions is low, especially when the nanoparticles are sparsely distributed. Therefore, melt-flow-driven effects are the dominant mechanism, and the effects from nanoparticles' size (radius) and density would be significant.

In Figure 4c_1–c_{12},d_1–c_6, nanoparticles with four different radii and two compositions are labeled uniformly with respect to their corresponding parent particles, in which ones with evident changes in their trajectories and position indicators are screened out. Comparing A1, A2 among Figure 4c_1–c_3; and B10, B15 among Figure 4c_5–c_8,d_3,d_4, a higher tendency to float is presented when increasing the size of nanoparticles of identical composition, or using Y_2O_3 than YIG for nanoparticles of identical size. Pair A1–A2 between Figure 4c_3,c_4 and pair C41–C44 between Figure 4c_9–c_{12}, on the other hand, indicate the potential interaction between particles. In particular, the latter pair presents a merged trajectory after one abrupt redirection, implying a potential agglomeration due to a short-range VDW attraction. A similar effect is also spotted for the case with relatively higher density (i.e., YIG) by comparing again the pair A1–A2 between Figures 4d_1,d_2 and pair C41–C42 between Figures 4d_5,d_6, demonstrating the potential enhanced interaction in an increased size for nanoparticles of identical composition/using YIG compared to Y_2O_3 for nanoparticles of identical size. Apart from these effects, a more complicated pattern variation in the overall trajectories change shall be discussed with respect to the change in nanoparticle size/composition. Unfortunately, information for current simulations (noting the nanoparticle are sparsely distributed) remains insufficient to deduct such a pattern change, which will be sufficiently covered in our upcoming works.

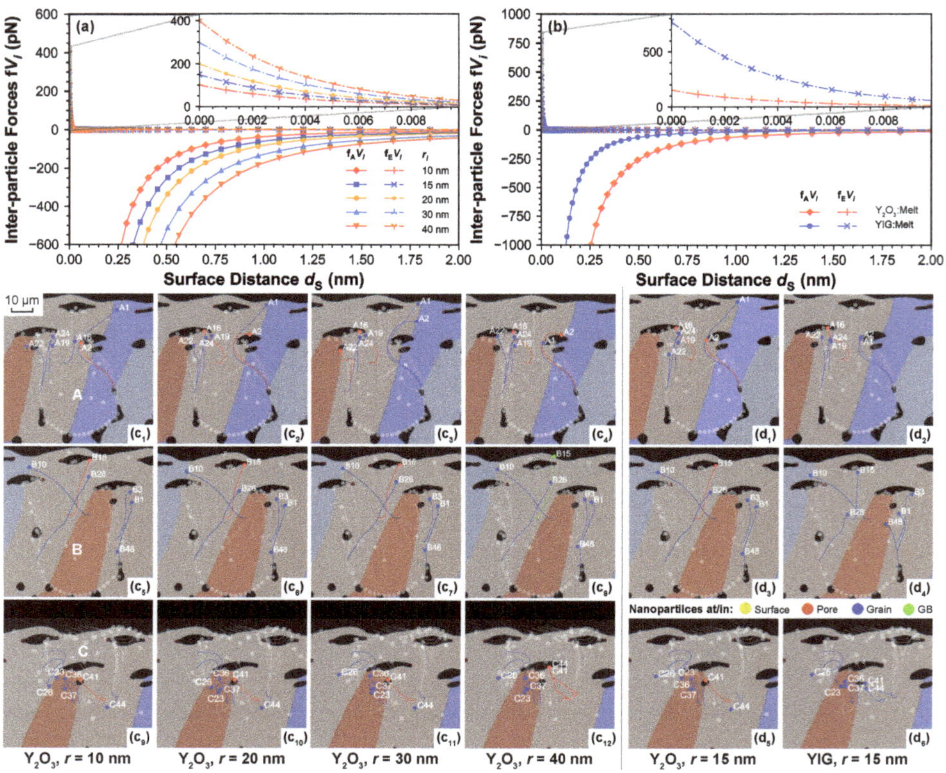

Figure 4. Diagrams of interactive forces between two nanoparticles of the same size vs. surface distance for cases (**a**) varying size (radius) for two Y_2O_3 nanoparticles, and (**b**) comparing nanoparticle compositions of Y_2O_3 and YIG for two nanoparticles with ri = 15 nm. Trajectories and relative positions at the final timestep are respectively for nanoparticles with varying radius and composition on different parent particle, i.e., for Y_2O_3 nanoparticles on parent particle A with varying radius of (**c$_1$**) 10 nm, (**c$_2$**) 20 nm, (**c$_3$**) 30 nm, (**c$_4$**) 40 nm; on parent particle B with radius of (**c$_5$**) 10 nm, (**c$_6$**) 20 nm, (**c$_7$**) 30 nm, (**c$_8$**) 40 nm; on parent particle C with radius of (**c$_9$**) 10 nm, (**c$_{10}$**) 20 nm, (**c$_{11}$**) 30 nm, (**c$_{12}$**) 40 nm; and for nanoparticles with monomodal radius of 15 nm on parent particle A with composition (**d$_1$**) Y_2O_3, (**d$_2$**) YIG; on parent particle B with composition (**d$_3$**) Y_2O_3, (**d$_4$**) YIG; on parent particle C with composition (**d$_5$**) Y_2O_3, (**d$_6$**) YIG. it demonstrates that the pattern of trajectories changes due to varying size/composition. By comparing the selected trajectories (labeled uniformly), an enhanced nanoparticle floating is observed when increasing the size of nanoparticles.

A statistic investigation of the nanoparticle evolution during processing is conducted to deduct features such as enrichment and agglomeration. The simulations shown in Figure 5 are performed for PM2000 powder with nine selected parent particles decorated with 0.005 wt % Y_2O_3 nanoparticles due to the limited computational efficiency, as presented in the inset of Figure 5a. Notice here that the weight percentage is calculated adapting the 2D scenario, i.e., wt % $= \frac{N_p \varrho_i \bar{r}_i^2}{\varrho_m R_I^2}$ with the total amount of nanoparticle N_p. In this sense, nanoparticles of $N_p = 1240$ are traced simultaneously. Three types of nanoparticle size distributions are investigated: the normal and the log-normal fittings of the large nanoparticle fraction measured in [23] and a monomodal obtained from the fitted mean radius $\bar{r}_i = 13$ nm of the distribution. To reduce the contamination to the statistics of the drift destinations, alloy powders located lower than the melt pool depth are not decorated with nanoparticles (inset in Figure 5a). Figure 5b presents the statistics of the drift destinations of the Y_2O_3 nanoparticles tested with different size distributions. Employing the adjacency test, in which the center distances (d_C) between nanoparticles are individually compared with the corresponding sum of mutual radii ($r_i + r_j$), the agglomerated nanoparticles can be identified. Following this criterion, the nanoparticles are classified as agglomerated/non-agglomerated, showing the corresponding fraction in Figure 5b. The results show that $\sim$ 55% of the nanoparticles are captured inside the grains after the LPBF process. Comparing different size distributions, monomodal and log-normal types have almost the same amount of in-grain nanoparticles (nano-inclusions), while the normal type shares the same agglomeration ratio as the log-normal type. However, only $\sim$ 2% of the nanoparticles end up in the grain boundaries. Within the nanoparticles in the grain boundaries, the log-normal distribution presents more agglomerated nanoparticles, reaching 46% (0.6% out of in-total 1.3%). In addition, there are nanoparticles that end up in the pores, $\sim$ 30%, and the surface, $\sim$13%. It is worth noting that these statistical results with a high fraction of in-grain nano-inclusions and a very low fraction of GB-captured ones may be attributed to the lack of driving effects from the migration of various interfaces, such as melt-grain interface and grain boundaries, which might force some of the nanoparticles to translate towards the migrating directions, pushing them away from the grain interior and eventually capturing them after the fusion of the powder bed.

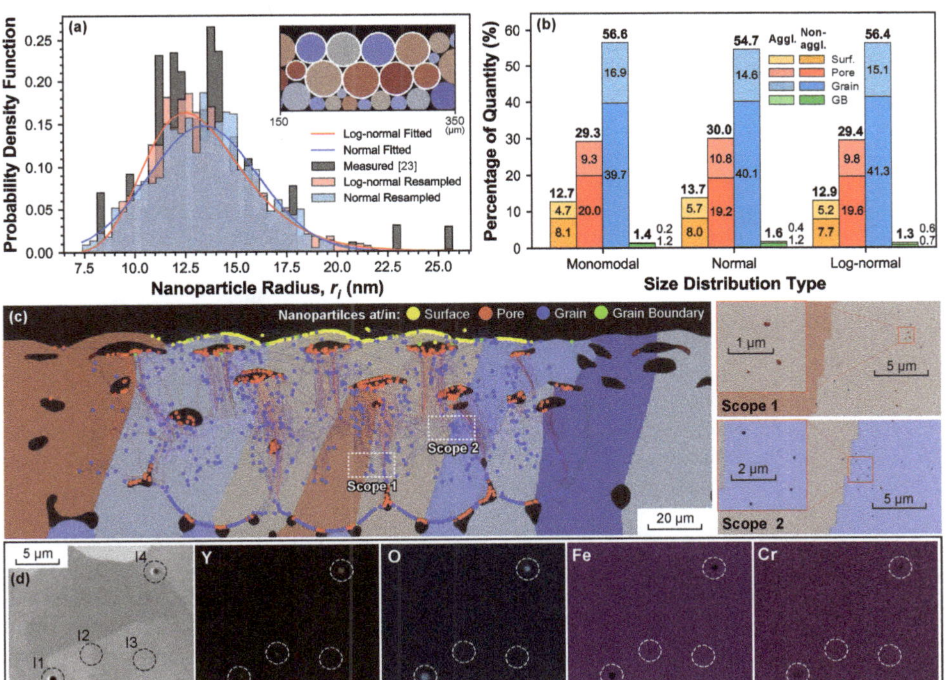

Figure 5. (**a**) Size distributions of nanoparticles. Inset: selected parent particles. (**b**) Statistics of drift destinations of nanoparticles with respect to distinct size distributions, presented as the percentage of quantity to the total amount of decorated nanoparticles, i.e., $N_p = 1240$, with the ones of agglomerated/non-agglomerated nanoparticles also presented as the component for each destination genre; (**c**) dispersion of traced Y_2O_3 nanoparticles with the log-normal size distribution in the LPBF-processed PM2000 polycrystal matrix (left) and regional magnification of dispersed nanoparticles in selected scopes (Scope 1 and 2) in realistic scale (right). Magnified scopes are employed to distinguish the occurrence of agglomeration from the enrichment, which is inaccessible in the trajectory illustration on the left. (**d**) SEM backscatter electron imaging with the corresponding EDX maps of the elemental Y, O, Fe, and Cr content in four inclusions (I1-I4) for identifying nanoparticle agglomeration. I3 is not agglomerated nanoparticles due to a lack of Y content, while I1 implies agglomerated nanoparticles on the surface of another oxide inclusion. The magnified scope in (**c**) right is also compared to the SEM imaging in (**d**), demonstrating relatively larger nanoparticle agglomeration (I4) spotted in experimental observation, while I2 shows almost a consistent scale to the simulated results.

In addition to the results provided in Figure 5b, the nanoparticle evolution during processing is graphically displayed in Figure 5c. This methodology is not only useful to predict the nanoparticle agglomeration but can be also extended to study the nanoparticle capture, enrichment, and agglomeration. However, even the evaluation of the agglomeration requires attentions on not only the trajectories overlapping but also the vast differences in spatial scales between the microstructure and the nanoparticles. For instance, Figure 5c presents the trajectories and positions of nanoparticles (log-normal type) dispersed in the LPBF-processed PM2000 polycrystalline matrix, in which two scopes are taken and magnified to visualize the nanoparticle scale. If only the trajectories are evaluated, one might prematurely conclude that Scope 2 has profound agglomerations comparing to Scope 1. However, after a magnified look, agglomerated nanoparticles only appear in Scope 1, while Scope 2 only contains locally enriched isolated nanoparticles.

The issue of identifying the nanoparticle agglomeration/enrichment should be also addressed in the experimental observation, yet the large spatial scale difference makes it difficult to characterize both the steel microstructure and nanoparticle size and dispersion in a single measurement. An example of an experimental SEM-EDX measurement from an LPBF processed sample of PM2000 steel decorated with 0.08 wt % Y_2O_3 is presented in Figure 5d. Four inclusions (labeled as I1-I4) that can be initially thought of as nanoparticle agglomerations during processing can be observed [12]. However, to evaluate whether the inclusions are pores or agglomerated nanoparticles, an elemental mapping with techniques such as EDX is required. The EDX maps show that, while no Y or O content is found in I3, these elements are present in I1, I2, and I4, and so it can be concluded that they are nanoparticle agglomerated structures. Furthermore, there is solely a small fraction of Y content found on the surrounding of I3 while O content is fully presented in the interior, implying agglomerated Y_2O_3 nanoparticles on the surface of another oxide inclusion that might be attributed to impurities. Another interesting point is that the diameter of some nanoparticle agglomerations spotted in Scope 1 of Figure 5c are less than 1 μm, consisting of merely countable nanoparticles, while the ones (esp. I4 in Figure 5d) experimentally observed are larger, and even much larger if compared with the original size of the Y_2O_3 nanoparticles—even though there are some (e.g., I2 in Figure 5d) showing consistency in scale. A possible reason is the relatively low mass fraction of the additivated nanoparticles in the tracing simulation (0.005%) compared to the experimental one (0.08%), which increases the probability of the nanoparticle interactions leading to agglomeration during processing. In future steps, the nanoparticle concentration employed in the simulation will be increased to match the experimental conditions.

5. Conclusions

In this work, we proposed a simulation scheme joining the heat-melt-microstructure-coupled phase-field model and the nanoparticle kinematics to trace nanoparticle during the LPBF process of the ODS steels, which is experimentally inaccessible. Simulations on stable LPBF single scan of a ferritic PM2000 nanoparticle-additivated powder bed were conducted for factors such as the nanoparticle composition and size distribution. The following conclusions can be drawn from this combined numerical and experimental study:

1. Simulation results provide the chronological location and located phases of the traced nanoparticle. This helps to depict nanoparticle drift associated with the evolution of local melt flow as well as the morphology, such as migrating nanoparticles driven by melt-flow or carried by floating/deforming bubbles, or stationary ones in the melt pool bottom area. Events such as nanoparticle capture (by grain/pore/grain boundary), enrichment, and potential agglomeration can be also visualized via trajectories and position indicator.
2. The drift and interactions of nanoparticles with different sizes and compositions (Y_2O_3 and YIG) are analyzed. By comparing the trajectories and positions of selected nanoparticles (or nanoparticle pairs) among cases, some preliminary discussions can be conducted regarding the influences of the nanoparticle size and compositions.

An enhanced nanoparticle floating is observed when the size is increased for nanoparticles of identical composition or when using Y_2O_3 for nanoparticles of identical size. In addition, an enhanced nanoparticle interaction is observed when increasing the size of nanoparticles of identical composition or using YIG for nanoparticles of identical size. Note that the above conclusions are made under the condition that nanoparticles are sparsely decorated (i.e., nanoparticles are sufficiently spaced). In this scenario, the melt-flow-driven effect is expected to dominate the process.

3. LPBF simulations of a PM2000 powder bed are conducted, in which nine parent particles are additivated with 0.005% Y_2O_3 nanoparticles. Three size distributions are evaluated i.e., monomodal, normal, and log-normal distributions. The results show that $\sim$ 55% of the nanoparticles are eventually captured by a grain, while merely $\sim$ 2% ones end up in the grain boundaries. Although the differences of nanoparticle location for the different size distributions are small, the monomodal case presents a relatively higher agglomeration ratio in grain-captured nanoparticles (nano-inclusions), while the log-normal type shows a higher agglomeration ratio in GB-captured nanoparticles.
4. By visualizing the traced nanoparticles on a nanometric scale, nanoparticle agglomeration and enrichment spotted in the simulation are distinguished. Comparisons between the simulations and experimental results show promising similarities, proving the potential of the simulation methodology to optimize the LPBF process parameters in order to reduce agglomeration effects and maximize the material reinforcement achieved in ODS steels.

The proposed scheme and models should be further extended in the near future for different aspects, e.g., implementation of nanoparticle tracing code in a computationally-efficient way to enable decoration with a larger mass fraction of the nanoparticles. Further effects such as the driving forces from migrating interfaces, and interactions between nanoparticles and the surface of the parent particles (as ideally the semi-infinity large interfaces) should also be considered.

Author Contributions: Conceptualization, B.-X.X., Y.Y., and B.G.; methodology, Y.Y. and B.-X.X.; software, Y.Y.; validation, Y.Y. and C.D.-B.; investigation, Y.Y., C.D.-B.; formal analysis, Y.Y., C.D.-B., and T.D.O.; resources, B.G. and C.D.-B.; data curation, Y.Y.; writing—original draft preparation, Y.Y. and C.D.-B.; writing—review and editing, Y.Y., C.D.-B., T.D.O., B.G., and B.-X.X.; visualization, Y.Y.; supervision, B.-X.X.; funding acquisition, B.G. and B.-X.X. All authors have read and agreed to the published version of the manuscript.

Funding: Authors acknowledge the financial support of the German Research Foundation (DFG) under the Collaborative Research Centre Transregio 270 (CRC-TRR 270, project number 405553726, sub-projects A06, B08 and Z01), the Research Training Groups 2561 (GRK 2561, project number 413956820, sub-project A4), and the Priority Programs 2256 (SPP 2256, project number 441153493) and 2122 (SPP 2122, project number 359962234). B.G. additionally acknowledges funding from the DFG within the Heisenberg Program, project number 445127149 (GO 2566/10-1).

Institutional Review Board Statement: Not applicable.

Informed Consent Statement: Not applicable.

Data Availability Statement: The authors declare that the data supporting the findings of this study are available within the paper.

Acknowledgments: The authors acknowledge support by the German Research Foundation (DFG) and the Open Access Publishing Fund of Technische Universitä Darmstadt. The authors greatly appreciate their access to the Lichtenberg High-Performance Computer and the technique supports from the HHLR, Technische Universitä Darmstadt. The authors highly thank Baptiste Gault and Philipp Kürnsteiner for the SEM-EDS measurements performed at the Max-Planck-Institut für Eisenforschung in Düsseldorf (MPIE).

Conflicts of Interest: The authors declare no conflict of interest.

Abbreviations

The following abbreviations are used in this manuscript:

BC	Boundary condition
DED	Direct energy deposition
EDX	Energy dispersive X-ray spectroscopy
GB	Grain boundary
HMM	Heat-melt-microstructure
IC	Initial condition
LAM	Laser-based additive manufacturing
LPBF	Laser powder bed fusion
NSAC	Navier-Stokes-Allen-Cahn
NSCH	Navier-Stokes-Cahn-Hilliard
ODS	Oxide dispersion strengthened
OP	Order parameter
PJFNK	Preconditioned Jacobian-Free Newton-Krylov
SEM	Scanning electron microscope
VDW	Van der Waals
YIG	Yttrium iron garnet

Symbols

The following symbols are used in this manuscript:

ρ	Conserved order parameter indicating the substance and pores/atmosphere
ϕ_S, ϕ_L, ϕ	Non-conserved order parameters indicating the solid (ϕ_S) and liquid (ϕ_L) phase. Due to the substance constraint $\phi_L = \rho - \phi_S$, only one ϕ is sufficient to represent the solid substance and $(\rho - \phi)$ to represent liquid substance
$\{\eta_\alpha\}$	Non-conserved order parameters indicating grain orientations
$\mathbf{u}$	Velocity field of melt flow
p	Hydrodynamic pressure
T	Temperature field
T_M, T_E, T_P	Melting temperature, environmental temperature, and pre-heating temperature, respectively
$f_{loc}, f_{grad}, f_{ht}$	Local, gradient and thermal terms of Helmholtz free energy density
$\underline{A}, \underline{B}, \underline{G}$	Model parameters with no dimension
$\underline{C}_{pt}, \underline{D}_{pt}, \underline{H}_{pt}$	Model parameters with dimension of free energy density
$\underline{C}_{cf}, \underline{D}_{cf}, \underline{H}_{cf}$	Model parameters with dimension of entropy density
$\kappa_\rho, \kappa_\phi, \kappa_\eta$	Gradient constants of corresponding order parameters
$\Phi_{ss}, \Phi_{at}, \Phi_{sf}, \Phi_{gb}, \Phi_S, \Phi_L$	Interpolation functions with subscript 'ss' standing for the substance, 'at' for the atmosphere/pore, 'sf' for the surface, 'gb' for the grain boundary, 'S' for the solid and 'L' for the liquid/melt
$P_l, \mathbf{v}_l, v_l$	Laser power, scan velocity and speed, respectively. $v_l = \lvert\mathbf{v}_l\rvert$
R_l	Nominal laser radius
q_v	Internal heat source for modeling the laser-induced thermal effect
β	Attenuation coefficient of the powder bed
ζ	Characteristic penetration depth of the powder bed
h	Heat convective coefficient
k_S, k_L, k_{at}	Heat conductivity coefficients with subscripts indicating the corresponding phases

ν_S, ν_L, ν_{at}	Dynamic viscosity coefficients with subscripts indicating the corresponding phases
$\varrho_S, \varrho_L, \varrho_{at}$	Densities with subscripts indicating the corresponding phases
c_S^p, c_L^p, c_{at}^p	Specific enthalphies with subscripts indicating the corresponding phases
c_r	Relative volumetric specific heat
$\mathscr{L}$	Latent heat during melting/solidification
G_{gb}^{eff}	Grain boundary mobility
$D_{ss}, D_{at}, D_{sf}, D_{gb}$	Self-diffusivity coefficients with subscripts indicating the corresponding phases
M	Isotropic Cahn–Hilliard mobility
L_ϕ, L_η	Isotropic Allen–Cahn mobilities
b	Magnitude of the resultant body forces acting on the melt flow
$\mathrm{Re}, \mathrm{Ste}, \mathrm{Pe}_T, \mathrm{Pe}_\rho, \mathrm{Ac}_\phi, \mathrm{Ac}_\eta$	Reynold number, Froude number, Stefan number, Péclet numbers for thermal and mass transfer, Allen–Cahn numbers for melting-solidification and grain growth, respectively
$V_i, \varrho_i, \mathbf{v}_i$	Volume, density and translation velocity of the i-th dispersed rigid nanoparticle in the melt respectively
$\mathbf{f}_M, \mathbf{f}_A, \mathbf{f}_E$	Driving force densities due to the melt flow, the VDW and the electrostatic interaction, respectively
σ	Cauchy stress tensor of the melt
d_C, d_S	Inter-particle center distance and surface distance, respectively
U_A, U_E	Inter-particle potential of VDW and the electrostatic interactions, respectively
$A_{p:v}, A_{m:v}, A_{p:m}$	Hamaker coefficients with subscript 'p:v' standing for nanoparticle in the vacuum, 'm:v' for medium/melt in the vacuum, and 'p:m' for nanoparticle in the medium/melt
Z	Screened-Coulombic interaction coefficient
ψ_i	Nanoparticle surface potential
$\varepsilon_p, \varepsilon_m, \varepsilon_0$	Permittivities of the nanoparticle, medium and vacuum, respectively
N_p	Total amount of the nanoparticle,
n_p, n_m	Refractive indices of the nanoparticle and the medium, respectively
λ_P	Plasma Debye length
n_e, ω_e	Free electron density and electronic absorption frequency of the melt
z_l, c_l, m_l	Valence electron number, atom fraction and the atom mass of the l-th element in the alloy, respectively
$k_B, \hbar, e_0, N_A$	Boltzmann constant, reduced Planck constant, elementary charge, and Avogadro constant, respectively
$\mathbb{V}(\mathbf{X}_I, R_I)$	Powder bed as a set, containing the center coordinates ($\mathbf{X}_I$) and the radius (R_I) of I-th powder
∇, ∇_d	Gradient operators. Subscript 'd' represents the gradient along the inter-particle direction

References

1. Martin, J.H.; Yahata, B.D.; Hundley, J.M.; Mayer, J.A.; Schaedler, T.A.; Pollock, T.M. 3D printing of high-strength aluminium alloys. *Nature* **2017**, *549*, 365–369. [CrossRef] [PubMed]
2. Kürnsteiner, P.; Wilms, M.B.; Weisheit, A.; Gault, B.; Jägle, E.A.; Raabe, D. High-strength Damascus steel by additive manufacturing. *Nature* **2020**, *582*, 515–519. [CrossRef] [PubMed]
3. Gu, D.; Shi, X.; Poprawe, R.; Bourell, D.L.; Setchi, R.; Zhu, J. Material-structure-performance integrated laser-metal additive manufacturing. *Science* **2021**, *372*. [CrossRef]

4. Orowan, E. Discussion on internal stresses. In *Symposium on Internal Stresses in Metals and Alloys*; Institute of Metals London: London, UK , 1948; pp. 451–453.
5. Wharry, J.P.; Swenson, M.J.; Yano, K.H. A review of the irradiation evolution of dispersed oxide nanoparticles in the bcc Fe-Cr system: Current understanding and future directions. *J. Nucl. Mater.* **2017**, *486*, 11–20. [CrossRef]
6. Jiang, Y.; Smith, J.R.; Odette, G.R. Prediction of structural, electronic and elastic properties of Y2Ti2O7 and Y2TiO5. *Acta Mater.* **2010**, *58*, 1536–1543. [CrossRef]
7. Suresh, K.; Nagini, M.; Vijay, R.; Ramakrishna, M.; Gundakaram, R.C.; Reddy, A.; Sundararajan, G. Microstructural studies of oxide dispersion strengthened austenitic steels. *Mater. Des.* **2016**, *110*, 519–525. [CrossRef]
8. Schneibel, J.; Heilmaier, M.; Blum, W.; Hasemann, G.; Shanmugasundaram, T. Temperature dependence of the strength of fine-and ultrafine-grained materials. *Acta Mater.* **2011**, *59*, 1300–1308. [CrossRef]
9. Unocic, K.A.; Pint, B.A.; Hoelzer, D.T. Advanced TEM characterization of oxide nanoparticles in ODS Fe–12Cr–5Al alloys. *J. Mater. Sci.* **2016**, *51*, 9190–9206. [CrossRef]
10. Karak, S.; Chudoba, T.; Witczak, Z.; Lojkowski, W.; Manna, I. Development of ultra high strength nano-Y2O3 dispersed ferritic steel by mechanical alloying and hot isostatic pressing. *Mater. Sci. Eng. A* **2011**, *528*, 7475–7483. [CrossRef]
11. Chang, H.J.; Cho, H.Y.; Kim, J.H. Stability of Y–Ti–O nanoparticles during laser melting of advanced oxide dispersion-strengthened steel powder. *J. Alloy. Compd.* **2015**, *653*, 528–533. [CrossRef]
12. Doñate-Buendia, C.; Kürnsteiner, P.; Stern, F.; Wilms, M.B.; Streubel, R.; Kusoglu, I.M.; Tenkamp, J.; Bruder, E.; Pirch, N.; Barcikowski, S.; et al. Microstructure formation and mechanical properties of ODS steels built by laser additive manufacturing of nanoparticle coated iron-chromium powders. *Acta Mater.* **2021**, *206*, 116566. [CrossRef]
13. Boegelein, T.; Louvis, E.; Dawson, K.; Tatlock, G.J.; Jones, A.R. Characterisation of a complex thin walled structure fabricated by selective laser melting using a ferritic oxide dispersion strengthened steel. *Mater. Charact.* **2016**, *112*, 30–40. [CrossRef]
14. Vasquez, E.; Giroux, P.F.; Lomello, F.; Nussbaum, M.; Maskrot, H.; Schuster, F.; Castany, P. Effect of powder characteristics on production of oxide dispersion strengthened Fe14Cr steel by laser powder bed fusion. *Powder Technol.* **2020**, *360*, 998–1005. [CrossRef]
15. Vasquez, E.; Giroux, P.F.; Lomello, F.; Chniouel, A.; Maskrot, H.; Schuster, F.; Castany, P. Elaboration of oxide dispersion strengthened Fe-14Cr stainless steel by selective laser melting. *J. Mater. Process. Technol.* **2019**, *267*, 403–413. [CrossRef]
16. Ghayoor, M.; Lee, K.; He, Y.; Chang, C.h.; Paul, B.K.; Pasebani, S. Selective laser melting of austenitic oxide dispersion strengthened steel: Processing, microstructural evolution and strengthening mechanisms. *Mater. Sci. Eng. A* **2020**, *788*, 139532. [CrossRef]
17. AlMangour, B.; Grzesiak, D.; Yang, J.M. Rapid fabrication of bulk-form TiB2/316L stainless steel nanocomposites with novel reinforcement architecture and improved performance by selective laser melting. *J. Alloy. Compd.* **2016**, *680*, 480–493. [CrossRef]
18. Brocq, M.; Radiguet, B.; Poissonnet, S.; Cuvilly, F.; Pareige, P.; Legendre, F. Nanoscale characterization and formation mechanism of nanoclusters in an ODS steel elaborated by reactive-inspired ball-milling and annealing. *J. Nucl. Mater.* **2011**, *409*, 80–85. [CrossRef]
19. Cayron, C.; Rath, E.; Chu, I.; Launois, S. Microstructural evolution of Y_2O_3 and $MgAl_2O_4$ ODS EUROFER steels during their elaboration by mechanical milling and hot isostatic pressing. *J. Nucl. Mater.* **2004**, *335*, 83–102. [CrossRef]
20. Bergner, F.; Hilger, I.; Virta, J.; Lagerbom, J.; Gerbeth, G.; Connolly, S.; Hong, Z.; Grant, P.S.; Weissgärber, T. Alternative Fabrication Routes toward Oxide-Dispersion-Strengthened Steels and Model Alloys. *Metall. Mater. Trans. A* **2016**, *47*, 5313–5324. [CrossRef]
21. Smith, T.M.; Thompson, A.C.; Gabb, T.P.; Bowman, C.L.; Kantzos, C.A. Efficient production of a high-performance dispersion strengthened, multi-principal element alloy. *Sci. Rep.* **2020**, *10*, 1–9. [CrossRef] [PubMed]
22. Moghadasi, M.A.; Nili-Ahmadabadi, M.; Forghani, F.; Kim, H.S. Development of an oxide-dispersion-strengthened steel by introducing oxygen carrier compound into the melt aided by a general thermodynamic model. *Sci. Rep.* **2016**, *6*, 38621. [CrossRef]
23. Doñate-Buendía, C.; Frömel, F.; Wilms, M.B.; Streubel, R.; Tenkamp, J.; Hupfeld, T.; Nachev, M.; Gökce, E.; Weisheit, A.; Barcikowski, S.; et al. Oxide dispersion-strengthened alloys generated by laser metal deposition of laser-generated nanoparticle-metal powder composites. *Mater. Des.* **2018**. [CrossRef]
24. Streubel, R.; Wilms, M.B.; Doñate-Buendía, C.; Weisheit, A.; Barcikowski, S.; Schleifenbaum, J.H.; Gökce, B. Depositing laser-generated nanoparticles on powders for additive manufacturing of oxide dispersed strengthened alloy parts via laser metal deposition. *Jpn. J. Appl. Phys.* **2018**, *57*, 040310. [CrossRef]
25. Springer, H.; Baron, C.; Szczepaniak, A.; Jägle, E.A.; Wilms, M.B.; Weisheit, A.; Raabe, D. Efficient additive manufacturing production of oxide- and nitride-dispersion-strengthened materials through atmospheric reactions in liquid metal deposition. *Mater. Des.* **2016**, *111*, 60–69. [CrossRef]
26. Williams, C.A.; Unifantowicz, P.; Baluc, N.; Smith, G.D.; Marquis, E.A. The formation and evolution of oxide particles in oxide-dispersion- strengthened ferritic steels during processing. *Acta Mater.* **2013**, *61*, 2219–2235. [CrossRef]
27. Li, S.; Xiao, H.; Liu, K.; Xiao, W.; Li, Y.; Han, X.; Mazumder, J.; Song, L. Melt-pool motion, temperature variation and dendritic morphology of Inconel 718 during pulsed- and continuous-wave laser additive manufacturing: A comparative study. *Mater. Des.* **2017**, *119*, 351–360. [CrossRef]
28. AlMangour, B.; Baek, M.S.; Grzesiak, D.; Lee, K.A. Strengthening of stainless steel by titanium carbide addition and grain refinement during selective laser melting. *Mater. Sci. Eng. A* **2018**, *712*, 812–818. [CrossRef]

29. Xu, J.Q.; Chen, L.Y.; Choi, H.; Li, X.C. Theoretical study and pathways for nanoparticle capture during solidification of metal melt. *J. Phys. Condens. Matter* **2012**, *24*. [CrossRef]
30. Heiden, M.J.; Deibler, L.A.; Rodelas, J.M.; Koepke, J.R.; Tung, D.J.; Saiz, D.J.; Jared, B.H. Evolution of 316L stainless steel feedstock due to laser powder bed fusion process. *Addit. Manuf.* **2019**, *25*, 84–103. [CrossRef]
31. King, W.E.; Anderson, A.T.; Ferencz, R.M.; Hodge, N.E.; Kamath, C.; Khairallah, S.A.; Rubenchik, A.M. Laser powder bed fusion additive manufacturing of metals; physics, computational, and materials challenges. *Appl. Phys. Rev.* **2015**, *2*, 041304. [CrossRef]
32. Zhang, D.; Qiu, D.; Gibson, M.A.; Zheng, Y.; Fraser, H.L.; StJohn, D.H.; Easton, M.A. Additive manufacturing of ultrafine-grained high-strength titanium alloys. *Nature* **2019**, *576*, 91–95. [CrossRef]
33. Khairallah, S.A.; Anderson, A. Mesoscopic simulation model of selective laser melting of stainless steel powder. *J. Mater. Process. Technol.* **2014**, *214*, 2627–2636. [CrossRef]
34. McCallen, C. *Technical Report LLNL-ABS-565212*; Lawrence Livemore National Laboratory: Livermore, CA, USA, 2012.
35. Shi, R.; Khairallah, S.A.; Roehling, T.T.; Heo, T.W.; McKeown, J.T.; Matthews, M.J. Microstructural control in metal laser powder bed fusion additive manufacturing using laser beam shaping strategy. *Acta Mater.* **2020**, *184*, 284–305. [CrossRef]
36. Khairallah, S.A.; Anderson, A.T.; Rubenchik, A.; King, W.E. Laser powder-bed fusion additive manufacturing: Physics of complex melt flow and formation mechanisms of pores, spatter, and denudation zones. *Acta Mater.* **2016**, *108*, 36–45. [CrossRef]
37. Körner, C.; Attar, E.; Heinl, P. Mesoscopic simulation of selective beam melting processes. *J. Mater. Process. Technol.* **2011**, *211*, 978–987. [CrossRef]
38. Attar, E.; Körner, C. Lattice Boltzmann method for dynamic wetting problems. *J. Colloid Interface Sci.* **2009**, *335*, 84–93. [CrossRef]
39. Attar, E.; Körner, C. Lattice Boltzmann model for thermal free surface flows with liquid–solid phase transition. *Int. J. Heat Fluid Flow* **2011**, *32*, 156–163. [CrossRef]
40. Rai, A.; Markl, M.; Körner, C. A coupled Cellular Automaton–Lattice Boltzmann model for grain structure simulation during additive manufacturing. *Comput. Mater. Sci.* **2016**, *124*, 37–48. [CrossRef]
41. Rai, A.; Helmer, H.; Körner, C. Simulation of grain structure evolution during powder bed based additive manufacturing. *Addit. Manuf.* **2017**, *13*, 124–134. [CrossRef]
42. Lian, Y.; Lin, S.; Yan, W.; Liu, W.K.; Wagner, G.J. A parallelized three-dimensional cellular automaton model for grain growth during additive manufacturing. *Comput. Mech.* **2018**, *61*, 543–558. [CrossRef]
43. Rolchigo, L. Application of alloy solidification theory to cellular automata modeling of near–rapid constrained solidification. *Comput. Mater. Sci.* **2019**, *163*, 148–161. [CrossRef]
44. Zhang, Y.; Zhang, J. Modeling of solidification microstructure evolution in laser powder bed fusion fabricated 316L stainless steel using combined computational fluid dynamics and cellular automata. *Addit. Manuf.* **2019**, *28*, 750. [CrossRef]
45. Yang, M.; Wang, L.; Yan, W. Phase-field modeling of grain evolutions in additive manufacturing from nucleation, growth, to coarsening. *Npj Comput. Mater.* **2021**, *7*, 56. [CrossRef]
46. Yang, Y.; Kühn, P.; Yi, M.; Egger, H.; Xu, B.X. Non-Isothermal Phase-Field Modeling of Heat–Melt–Microstructure-Coupled Processes during Powder Bed Fusion. *JOM* **2020**, *72*, 1719–1733. [CrossRef]
47. Xu, J. Achieving Uniform Nanoparticle Dispersion in Metal Matrix Nanocomposites. In *Doctor of Philosophy in Materials Science and Engineering*; University of California: Los Angeles, CA, USA, 2015.
48. Kamysbayev, V.; James, N.M.; Filatov, A.S.; Srivastava, V.; Anasori, B.; Jaeger, H.M.; Gogotsi, Y.; Talapin, D.V. Colloidal Gelation in Liquid Metals Enables Functional Nanocomposites of 2D Metal Carbides (MXenes) and Lightweight Metals. *ACS Nano* **2019**, *13*, 12415–12424. [CrossRef]
49. Paik, Y.H.; Yoon, W.J.; Shin, H.C. Static electrification of solid oxide in liquid metal and electrical double layer at the interface. *J. Colloid Interface Sci.* **2004**, *269*, 354–357. [CrossRef]
50. Paik, Y.H.; Pan, J.H.; Yoon, W.J. Charging phenomena at the metal oxide-liquid metal interfaces and determination of excess electron density of metal oxide-mercury systems by the induced emf method. *J. Colloid Interface Sci.* **2001**, *244*, 444–446. [CrossRef]
51. Zeng, M.; Cao, H.; Zhang, Q.; Gao, X.; Fu, L. Self-Assembly of Metal Oxide Nanoparticles in Liquid Metal toward Nucleation Control for Graphene Single-Crystal Arrays. *Chem* **2018**, *4*, 626–636. [CrossRef]
52. Yang, Y.; Oyedeji, T.D.; Kühn, P.; Xu, B.X. Investigation on Temperature-Gradient-Driven Effects in Unconventional Sintering *Via* Non-Isothermal Phase-Field Simulation. *Scr. Mater.* **2020**, *186*, 152–157. [CrossRef]
53. Yang, Y.; Ragnvaldsen, O.; Bai, Y.; Yi, M.; Xu, B.X. 3D Non-Isothermal Phase-Field Simulation of Microstructure Evolution during Selective Laser Sintering. *Npj Comput. Mater.* **2019**, *5*, 81. [CrossRef]
54. Lasers and laser-related equipment–Test methods for laser beam widths, divergence angles and beam propagation ratios–Part 1: Stigmatic and simple astigmatic beams. In *Standard ISO 11146-1: 2005 (E)*; International Organization for Standardization: Geneva, Switzerland , 2005.
55. Shukla, S.; Bhattacharjee, S.; Weber, A.Z.; Secanell, M. Experimental and Theoretical Analysis of Ink Dispersion Stability for Polymer Electrolyte Fuel Cell Applications. *J. Electrochem. Soc.* **2017**, *164*, F600–F609. [CrossRef]
56. Bergström, L.; Meurk, A.; Arwin, H.; Rowcliffe, D.J. Estimation of Hamaker constants of ceramic materials from optical data using Lifshitz theory. *J. Am. Ceram. Soc.* **1996**, *79*, 339–348. [CrossRef]
57. Jiang, L.; Rahnama, M.; Zhang, B.; Zhu, X.; Sui, P.C.; Ye, D.D.; Djilali, N. Predicting the interaction between nanoparticles in shear flow using lattice Boltzmann method and Derjaguin-Landau-Verwey-Overbeek (DLVO) theory. *Phys. Fluids* **2020**, *32*. [CrossRef]

58. Ladd, A.; Verberg, R. Lattice-Boltzmann simulations of particle-fluid suspensions. *J. Stat. Phys.* **2001**, *104*, 1191–1251.:1010414013942. [CrossRef]
59. Ouchene, R. Numerical simulation and modeling of the hydrodynamic forces and torque acting on individual oblate spheroids. *Phys. Fluids* **2020**, *32*, 073303. [CrossRef]
60. Glowinski, R.; Pan, T.W.; Hesla, T.I.; Joseph, D.D. A distributed Lagrange multiplier/fictitious domain method for particulate flows. *Int. J. Multiph. Flow* **1999**, *25*, 755–794. [CrossRef]
61. Glowinski, R.; Pan, T.W.; Hesla, T.I.; Joseph, D.D.; Périaux, J. A Fictitious Domain Approach to the Direct Numerical Simulation of Incompressible Viscous Flow past Moving Rigid Bodies: Application to Particulate Flow. *J. Comput. Phys.* **2001**, *169*, 363–426. [CrossRef]
62. Hamaker, H.C. The London-van der Waals attraction between spherical particles. *Physica* **1937**, *4*, 1058–1072. [CrossRef]
63. Lifshitz, E.M.; Hamermesh, M. The theory of molecular attractive forces between solids. In *Perspectives in Theoretical Physics*; Elsevier: Amsterdam, The Netherlands, 1992; pp. 329–349.
64. Dzyaloshinskii, I.E.; Lifshitz, E.M.; Pitaevskii, L.P. The general theory of van der Waals forces. *Adv. Phys.* **1961**, *10*, 165–209. [CrossRef]
65. Israelachvili, J.N. *Intermolecular and Surface Forces*; Academic press: Cambridge, MA, USA, 2011.
66. Ohshima, H. Electrostatic interaction between two spherical colloidal particles. *Adv. Colloid Interface Sci.* **1994**, *53*, 77–102. [CrossRef]
67. Tonks, M.R.; Gaston, D.; Millett, P.C.; Andrs, D.; Talbot, P. An object-oriented finite element framework for multiphysics phase field simulations. *Comput. Mater. Sci.* **2012**, *51*, 20–29. [CrossRef]
68. Elliott, C.M.; French, D.A.; Milner, F. A second order splitting method for the Cahn-Hilliard equation. *Numer. Math.* **1989**, *54*, 575–590. [CrossRef]
69. Balay, S.; Gropp, W.D.; McInnes, L.C.; Smith, B.F. Efficient management of parallelism in object-oriented numerical software libraries. In *Modern Software Tools for Scientific Computing*; Birkhäuser: Basel, Switzerland, 1997; pp. 163–202.
70. Peterson, J.W.; Lindsay, A.D.; Kong, F. Overview of the incompressible Navier–Stokes simulation capabilities in the MOOSE framework. *Adv. Eng. Softw.* **2018**, *119*, 68–92. [CrossRef]
71. Lu, L.X.; Sridhar, N.; Zhang, Y.W. Phase field simulation of powder bed-based additive manufacturing. *Acta Mater.* **2018**, *144*, 801–809. [CrossRef]
72. Permann, C.J.; Tonks, M.R.; Fromm, B.; Gaston, D.R. Order parameter re-mapping algorithm for 3D phase field model of grain growth using FEM. *Comput. Mater. Sci.* **2016**, *115*, 18–25. [CrossRef]
73. Moelans, N.; Blanpain, B.; Wollants, P. Quantitative analysis of grain boundary properties in a generalized phase field model for grain growth in anisotropic systems. *Phys. Rev. B* **2008**, *78*, 024113. [CrossRef]
74. Chockalingam, K.; Kouznetsova, V.G.; van der Sluis, O.; Geers, M.G. 2D Phase field modeling of sintering of silver nanoparticles. *Comput. Methods Appl. Mech. Eng.* **2016**, *312*, 492–508. [CrossRef]
75. Chang, K.; Meng, F.; Ge, F.; Zhao, G.; Du, S.; Huang, F. Theory-guided bottom-up design of the FeCrAl alloys as accident tolerant fuel cladding materials. *J. Nucl. Mater.* **2019**, *516*, 63–72. [CrossRef]
76. MatWeb. Schwarzkopf Plansee PM 2000 Grain Class 4 ODS Iron Alloy, 2021. Available online: www.matweb.com/search/datasheet_print.aspx?matguid=96b81a5874794095964dae30a476b069 (accessed on 24 March 2021).
77. Hill, A.; Wallach, E.R. Modelling solid-state diffusion bonding. *Acta Metall.* **1989**, *37*, 2425–2437. [CrossRef]
78. Price, A.; Holl, H.; Greenough, A. The surface energy and self diffusion coefficient of solid iron above 1350 C. *Acta Metall.* **1964**, *12*, 49–58. [CrossRef]
79. Capdevila, C.; Chen, Y.L.; Jones, A.R.; Bhadeshia, H.K. Grain boundary mobility in Fe-base oxide dispersion strengthened PM2000 alloy. *ISIJ Int.* **2003**, *43*, 777–783. [CrossRef]
80. Hoshino, T.; Mito, K.; Nagashima, A.; Miyata, M. Determination of the thermal conductivity of argon and nitrogen over a wide temperature range through data evaluation and shock-tube experiments. *Int. J. Thermophys.* **1986**, *7*, 647–662. [CrossRef]
81. Chase, M.W., Jr. NIST-JANAF thermochemical tables. *J. Phys. Chem. Ref. Data, Monogr.* **1998**, *9*. [CrossRef]
82. Liu, F.; Zhang, Q.; Zhou, W.; Zhao, J.; Chen, J. Micro scale 3D FEM simulation on thermal evolution within the porous structure in selective laser sintering. *J. Mater. Process. Technol.* **2012**, *212*, 2058–2065. [CrossRef]
83. Kamaeva, L.V.; Sterkhova, I.V.; Lad'yanov, V.I. Viscosity and supercooling of Fe-Cr ($\leq$40 at *Inorg. Mater.* **2012**, *48*, 318–324. [CrossRef]
84. Dawe, R.A.; Smith, E.B. Viscosities of the Inert Gases at High Temperatures. *J. Chem. Phys.* **1970**, *52*, 693. [CrossRef]

materials

MDPI

Article

Improved Process Efficiency in Laser-Based Powder Bed Fusion of Nanoparticle Coated Maraging Tool Steel Powder

Oliver Pannitz [1,*], Felix Großwendt [2], Arne Lüddecke [3], Arno Kwade [3], Arne Röttger [4] and Jan Torsten Sehrt [1]

1 Hybrid Additive Manufacturing, Ruhr University Bochum, Universitätsstr. 150, 44780 Bochum, Germany; jan.sehrt@rub.de
2 Materials Technology, Ruhr University Bochum, Universitätsstr. 150, 44780 Bochum, Germany; felix.grosswendt@rub.de
3 Institute for Particle Technology, Technical University Braunschweig, Volkmaroder Str. 5, 38104 Braunschweig, Germany; a.lueddecke@tu-bs.de (A.L.); a.kwade@tu-bs.de (A.K.)
4 Novel Manufacturing Technologies and Materials, University of Wuppertal, Bahnhofstr. 15, 42651 Solingen, Germany; roettger@uni-wuppertal.de
* Correspondence: oliver.pannitz@ruhr-uni-bochum.de; Tel.: +49-234-32-23519

Abstract: Research and development in the field of metal-based additive manufacturing are advancing steadily every year. In order to increase the efficiency of powder bed fusion of metals using a laser beam system (PBF LB/M), machine manufacturers have implemented extensive optimizations with regard to the laser systems and build volumes. However, the optimization of metallic powder materials using nanoparticle additives enables an additional improvement of the laser–material interaction. In this work, tool steel 1.2709 powder was coated with silicon carbide (SiC), few-layer graphene (FLG), and iron oxide black (IOB) on a nanometer scale. Subsequently, the feedstock material and the modified powder materials were analyzed concerning the reflectance of the laser radiation and processed by PBF-LB/M in a systematic and consistent procedure to evaluate the impact of the nano-additivation on the process efficiency and mechanical properties. As a result, an increased build rate is achieved, exhibiting a relative density of 99.9% for FLG/1.2709 due to a decreased reflectance of this modified powder material. Furthermore, FLG/1.2709 provides hardness values after precipitation hardening with only aging comparable to the original 1.2709 material and is higher than the SiC- and IOB-coated material. Additionally, the IOB coating tends to promote oxide-formation and lack-of-fusion defects.

Keywords: additive manufacturing; PBF-LB/M; tool steel (1.2709); nanocomposite; microstructure; mechanical properties

Citation: Pannitz, O.; Großwendt, F.; Lüddecke, A.; Kwade, A.; Röttger, A.; Sehrt, J.T. Improved Process Efficiency in Laser-Based Powder Bed Fusion of Nanoparticle Coated Maraging Tool Steel Powder. *Materials* **2021**, *14*, 3465. https://doi.org/10.3390/ma14133465

Academic Editor: Federico Mazzucato

Received: 1 June 2021
Accepted: 18 June 2021
Published: 22 June 2021

1. Introduction

Direct additive manufacturing of metallic components is ensured, among other things, by powder bed fusion of metals using a laser beam system (PBF-LB/M) [1]. In this process, a thin powder layer is applied across a build platform, and the calculated cross-sectional areas (slices) of the desired components are melted by the laser beam and permanently bonded to the underlying layer. These process steps are repeated until the components are complete [2]. The process efficiency of the PBF-LB/M process can be achieved, on the one hand, by adapting the additive manufacturing system itself and, on the other hand, by modifying the powder material to be processed. The former results in the use of PBF-LB/M systems equipped with larger build volumes, multiple laser beams, or high-power laser systems, also leading to a significant increase in equipment prices. The modification of commercially available powder feedstocks by nanoparticle additivation has been addressed multiple times in the scientific literature, as depicted in the following sections. This enables more efficient process control and can promote the processing of dense and defect-free components with improved mechanical properties.

1.1. PBF-LB/M Process of Composite Powder

In a review, Kusoglu et al. [3] used aluminum powder as an exemplary case to illustrate how nanoparticle modification has been the focus of research in recent years in order to specifically adapt materials to the PBF-LB/M process. The research group of Gu et al. [4–6] focused extensively on the additivation of Ti and AlSi10Mg powder with TiC nanoparticles. Initially, a relative density of TiC/Ti (12.5 wt.% TiC) of above 98.3% was achieved in PBF-LB/M. Furthermore, an increased microhardness of 577 $HV_{0.2}$ and decreased coefficient of friction (COF) of 0.19 were observed. As a comparison, the microhardness of PBF-LB/M-processed pure Ti showed a maximum hardness of 287 $HV_{0.2}$. A more increased TiC content did not lead to any further improvement in hardness and wear resistance due to the decreasing relative density. In addition, a change of the microstructure from a coarsened dendritic to a uniformly dispersed nanoscale lamellar microstructure was observed.

Nano-additivation by TiC-particles was also regarded in the research activities of Al Mangour et al. [7,8]. They investigated the influence of different TiC particle sizes in a stainless steel 1.4404 matrix. As a result, the addition of fine TiC particles improved the wettability and promoted the densification behavior of 1.4404 steel processed by PBF-LB/M. Furthermore, the combined effects of grain-refinement and grain-boundary strengthening enable an increased microhardness, coefficient of friction, and decreased wear to a simultaneously increased volume content of TiC [7]. In subsequent research activities from Al Mangour et al. [8], the thermal behavior during laser–material interaction, microstructural characteristics, and tribological performance of TiC/1.4404 was investigated in an experimental and simulative approach. Thereby, they reported that the Marangoni convection correlates with the applied volume energy density. An increased volume energy density also promotes a ring-like formation of TiC nanoparticles in the melt pool and prevents its aggregation [8]. TiB_2/1.4404 was also processed with varying nanoparticle content by Al Mangour et al. [9]. An increased hardness, yield strength, and decreased coefficient of friction and wear rates with an increased TiB_2 volume content could be observed. Further-reinforced 1.4404 powder processed by PBF-LB/M was investigated by Zhao et al. [10]. By decorating the surface of the stainless-steel powder 1.4404 with TiC (0 wt.%, 2 wt.%, 4 wt.%) nanoparticles a refinement of cell size was noticeable. Consequently, an increase in hardness by 12.4% could be observed by adding 2 wt.% TiC. The addition of a mass fraction of 4 wt.% results in excessive particles that impair the densification of the PBF-LB/M manufactured components. Doñate et al. [11] homogeneously modified the surface of an Fe-Cr alloy with Y_2O_3 nanoparticles using pulsed laser fragmentation in water. The powder was consolidated by PBF-LB/M and DED-LB/M (directed energy deposition of metals using a laser beam) and was characterized by an increased microhardness. Chang et al. [12] studied the influence of different particle sizes in a SiC/AlSi10Mg composite powder on a formed microstructure and the associated mechanical properties of PBF-LB/M parts. By using the smallest SiC particle size, dense and defect-free microstructures with a high microhardness of 218.5 $HV_{0.1}$ were achieved, which corresponds to an increase of 50% compared to the AlSi10Mg powder feedstock. In addition, Sehrt et al. [13] processed Al_2O_3 nanoparticles on feedstock material tool steel (1.2709) and Hastelloy X (2.4665) and investigated the melt pool dynamics and microstructural properties. Both nanocomposites show a significant improvement of mechanical properties. The nanoparticle-modified Hastelloy X exhibited a microhardness of up to 494.3 $HV_{0.2}$, which is 80.8% higher than the original feedstock material (273.3 $HV_{0.2}$). Similar results were obtained with the modified tool steel. Accordingly, an increase in microhardness from 371.8 $HV_{0.2}$ to 529.3 $HV_{0.2}$ could be observed. However, the melt pool dynamics of the modified powder materials during the PBF-LB/M process were disturbed, which could be depicted by small dark crater like structures on the specimen's surfaces and varying widths of the melt pools. Hence, the energy input was not sufficient to create a homogeneous interconnection between adjacent weld lines. A similar approach was performed by Sehrt et al. [14] by modifying tool steel with WC and TiO_2 nanoparticles. Furthermore, in order to explicitly adjust copper materials for the PBF-LB/M process, carbon nanoparticles were applied on the copper particles' surfaces in an extensive work by Jadhav et al. [15–17]. By applying 0.1 wt.% of carbon nanoparticles on the surface of pure copper particles, a relative density of 98% could be achieved while using more efficient process parameters due to an increased absorption. In addition, the flowability was improved compared to the virgin powder [15]. Comparable results could be achieved using 0.05 wt.% on pre-alloyed CuCr0.3 powder concerning the coupling of laser radiation. In addition, mechanical properties, such as tensile strength and YS, could be increased by 56 MPa and 79 MPa, respectively.

1.2. Laser Absorption of Metal Powders

The PBF-LB/M process is characterized by various influencing variables on the equipment side, on the material side, and with regard to the external environment, which impedes the exclusion of specific factors [18,19]. The absorption behavior of laser radiation from metallic powder materials in the PBF-LB/M process represents a decisive factor for the formation of a dense microstructure of the components. On the one hand, the absorption behavior depends on the exposure parameters and consequently, the duration of laser–material interaction as investigated by Fu et al. [20] presented with finite element analysis. On the other hand, the material itself represents a significant influencing factor. Usually, a laser with a wavelength of 1064 nm is used for PBF-LB/M, which is well suited for processing steels, aluminum alloys, nickel alloys, etc. Recently, a green laser featuring a wavelength of 515 nm was shown to be more efficient for processing pure copper powder, which can be traced back to an increased absorption of the laser irradiation of up to ten times in contrast to 1064 nm lasers [21,22]. In addition to the material used, laser absorption is determined by particle size distribution (PSD), morphology, surface roughness, and optical appearance (color). The correlation between PSD and absorption behavior has been studied in a simulative approach using the physical ray-tracing method by Yang et al. [23] and Boley et al. [24]. Both studies show that higher median particle sizes lead to a decrease in absorption, which is caused by multiple reflections of laser radiation within the bulk powder. These results were confirmed by Gu et al. [25], who investigated the absorption of AlSi10Mg powder which has been modified with SiC and TiB_2. An increasing absorption was observed with the addition of smaller particles. In a previous publication of this research group, the influence of coated stainless steel powders using SiC nanoparticles and few-layer graphene was investigated [26]. Consequently, increasing surface roughness and a macroscopically darker appearance of the metal powder could be observed. Increased beam traps formed which led to increased absorption. This result is in accordance with the work of Zhou et al. [27], who investigated the influence of Al_2O_3-coated metallic powder in terms of absorption. Furthermore, Gruber et al. [28] analyzed the absorption behavior of an Inconel 718 powder in four different conditions: virgin, used, overflow, and spatter. The latter was located at the inert gas outlet after the PBF-LB/M process and contained both process by-products and used powder. It was characterized by a significantly higher absorption value compared the virgin and used powder, which can also be attributed to a darker discoloration and modified surface quality of the powder particles due to oxide spots.

1.3. Determination of Process Parameters for PBF-LB/M

Standardized and systematic PBF-LB/M processes must be established and implemented to ensure reproducible and transferable parameter qualification for powder materials. Some researchers initiated parameter analyses by investigating the geometrical properties of single tracks (exposure of single scan vectors) with different energy intensities [29–33]. In addition to geometric properties, optical anomalies were also evaluated. Thus, a low energy density leads to irregular melting, poor wettability, and results in a partially spherical appearance of the melt. In this context, the term balling phenomenon is often used. In contrast, melt pools exhibiting an increased width are formed at increased energy inputs. According to Di et al. [29], a thin and homogeneous melt pool is preferred for a stable PBF-LB/M process. Consequently, qualitative aspects of single tracks are set up as a function of increasing energy densities:

- Irregular and pre-balling shape
- Regular but occasionally broken shape
- Regular and thin shape
- Regular and thick shape

In this work, tool steel 1.2709 powder was coated with silicon carbide (SiC), few-layer graphene (FLG), and iron oxide black (IOB). Subsequently, the powder feedstock and the modified powder materials were analyzed concerning their reflectance behavior and processed in a systematic and consistent procedure using PBF-LB/M to evaluate a change in process efficiency. Consideration of the relative density, hardness, and microstructure of the generated samples completes this study. In this contribution, the following scientific questions are addressed:

1. Does a change of surface properties of the metallic powder particles due to surface modification by nanoparticles lead to an increase in absorption? What is the reason for a change in absorption behavior?
2. An increased absorption rate indicates that more photons per time are introduced into the powder material. Does this simultaneously enable more efficient process control for the manufacturing of dense components/microstructures or do other influencing material properties have to be taken into account?
3. A systematic for qualifying exposure parameters was developed to ensure reproducibility and transferability. Can this system be used to manufacture dense components?
4. Both, the as-built and heat-treated specimens are analyzed regarding the microstructure and the hardness. What are the effects of the nanoparticles on the final part quality?

2. Materials and Methods

2.1. Feedstock Material

As feedstock material, inert gas atomized 1.2709 (X3NiCoMoTi18-9-5) tool steel powder was used. In addition, nanoparticle-coated 1.2709 powder was used. The applied nanoparticles were silicon carbide (SiC), few-layer graphene (FLG), and iron oxide black (IOB). The chemical composition of the tool steel 1.2709 is listed in Table 1.

Table 1. Chemical composition of the nominal tool steel 1.2709 in mass% and the tool steel used in this work.

Tool Steel	C	Ni	Co	Mo	Ti	Cr	Si	Mn	P	S	O	Fe
Nominal	≤0.03	17.00–19.00	8.50–10.00	4.50–5.20	0.80–1.20	≤0.25	≤0.10	≤0.15	≤0.01	≤0.01	-	bal.
Used powder	0.01	17.36	9.31	5.59	1.18	0.14	0.05	0.01	0.00	0.00	0.02	bal.

2.2. Additive Formulation and Coating Process of the Feedstock Material

The nano-particle SiC (E-SINSIC), FLG (C-NERGY KS6L), and IOB (Manganese ferrite black spinel, 48,447) were purchased from ESK-SIC GmbH (Frechen, Germany), Imerys Graphite & Carbon Switzerland Ltd., and Kremer Pigmente GmbH & Co. KG (Aichstetten, Germany), respectively. Each particle system was comminuted via stirred media milling to a particle size of around 100 nm for SiC and IOB. The graphite was delaminated under mild conditions in a stirred media mill to prevent lateral fractures and the chipping of edges. The resulting few-layer graphene (FLG) exhibited a lateral dimension ×50,3 of 568.3 nm. Consequently, the commercially available AM-feedstock powder tool steel 1.2709 (supplied by ThyssenKrupp Materials GmbH, Essen, Germany) was coated using a fluidized bed system (MiniGlatt, Glatt GmbH, Binzen, Germany) with the three produced suspensions. The IOB and SiC coating amounts were calculated in order to provide a single layer of nanoparticles covering the feedstock steel particles, resulting in a volume percentage of 1 vol.% of the added nanoparticles. For the FLG platelets, 0.75 vol.% was used for the coating to prevent an excess of FLG, which could lead to agglomeration and stacking of the platelets on the steel particle surfaces. Detailed information on the additive production and fluidized bed coating can be found in previous work [34].

2.3. *Laser Reflectance Measurement of the Feedstocks*

Diffuse reflectance infrared Fourier transform spectroscopy (DRIFTS) was performed with a Nicolet iS20 FTIR spectrometer from Thermo Fisher Scientific (Waltham, MA, USA), including the accessory DiffusIR from Pike Technologies (Madison, WI, USA) to quantify the reflectance of the different powder materials. The exact device specifications and measurement settings are presented in Table 2. A cup (diameter: 5 mm; depth: 3 mm) was prepared with the powder material to be measured. To imitate the powder distribution during the PBF-LB/M-process, the excess powder in the cup was removed by a coating device. Three spectra from each of the four samples were recorded to check the reproducibility of the measurements. A wavenumber of 9398 cm^{-1} (reciprocal value of the wavelength) corresponded to a wavelength of λ = 1064 nm (Nd:YAG laser in a PBF-LB/M-system).

Table 2. Settings for DRIFTS analysis.

Background Measurement	Aluminum Mirror
MCT/A detector range	4000 to 11,000 cm^{-1}
Scans per measurement	64
Spectral resolution	4 cm^{-1}
Environment	Room temperature
Further characteristics	XT-KBr beam splitter white light source

2.4. *Processing*

2.4.1. PBF-LB/M-System

The PBF-LB/M-system TruPrint 1000 of Trumpf GmbH & Co. KG (Ditzingen, Germany) was used to manufacture the specimens. Required properties regarding the laser system, coating mechanism, and inner atmosphere are shown in Table 3. Furthermore, a procedure was developed to ensure the reproducible and transferable qualifications of efficient PBF-LB/M parameters (Figure 1). The individual process steps are explained in more detail in the following sections.

Table 3. PBF-LB/M nominal process parameters and environmental properties applied to 1.2709, SiC/1.2709, FLG/1.2709, and IOB/1.2709.

Laser power, P_L	100–140 W
Hatch distance, h_d	Individual
Layer thickness, D_S	20 µm
Scan speed, v_S	600–1400 mm/s
Volume energy density, E_V	Individual
Scan strategy	90° alternating
Focal diameter	30 µm
Wavelength	1064 nm
Inert gas atmosphere	N_2
Gas flow rate	3 m/s
Recoating speed	80 mm/s
Coater type	Rubber x-profile

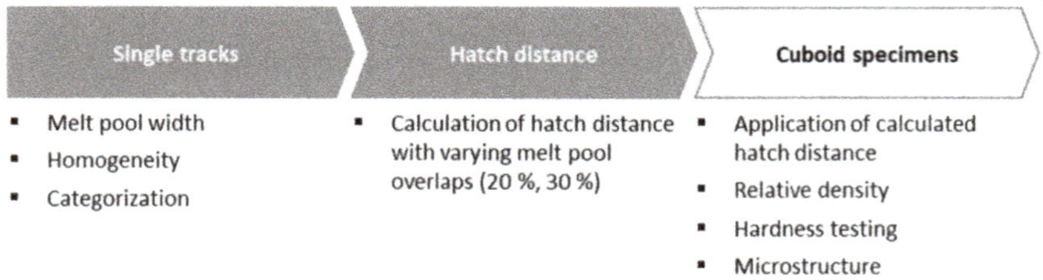

Figure 1. Process flow for determining PBF-LB/M-parameters.

2.4.2. Single Tracks

Individualized single tracks with a longitude of 5 mm were manufactured with a layer thickness of 0.02 mm. This procedure was repeated for 25 applied layers to avoid differences in heat transfer due to a non-heated build platform and differing diffusion processes due to a type C45 steel build platform material. In addition, the actual layer thickness was determined by the ratio of defined layer thickness and bulk density of the powder material and was achieved after 10 to 15 layers. In assessing the single tracks, the melt pool width and the homogeneity of the individual specimens were evaluated. Essential criteria for the evaluation of the specimens were conclusions about the consistency of the melt pool by evaluating and observation of the single uniform tracks. To quantify the melt pool width, it was measured at five equidistant positions along the complete length of a single track.

2.4.3. Cuboid Specimens

The three-dimensional specimens exhibited a size of $5 \times 5 \times 5$ mm^3 and were directly connected to the build platform without support structures. As illustrated in Figure 2a, a 90°-alternating scan strategy under a nitrogen gas atmosphere was applied. The layer thickness was kept constant at 0.02 mm, as depicted in Table 3. The scanning speed, laser power, and hatch distance varied and depended on the previous evaluation of the single tracks. According to literature research, an overlap of 20–30% of two adjacent welding lines is frequently used for the PBF-LB/M process [29,31].

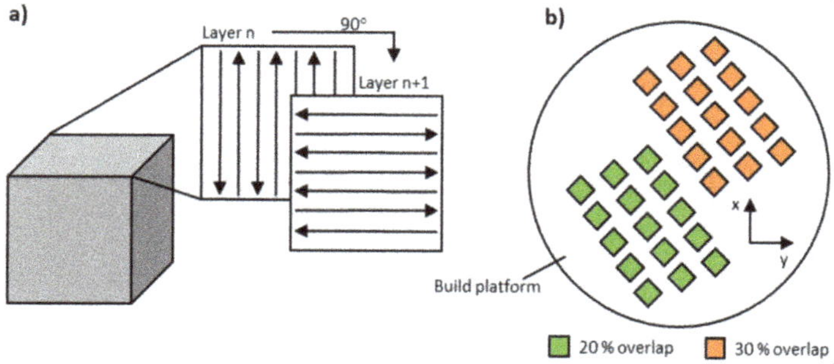

Figure 2. (**a**) A 3D model of the cuboid specimen with an illustration of the applied scan strategy. (**b**) Specimen arrangement on the build platform (top view).

2.4.4. Heat-Treatment

To achieve the desired application properties, maraging steels are typically solution-annealed and aged. To evaluate the influence of the nanoparticle additivation of the feedstock material, the formed microstructure and the associated mechanical properties of the additively manufactured specimens, different heat-treatments were carried out. On the one hand, a conventional heat treatment according to the manufacturer's specification was performed. The conventional heat-treatment included solution annealing at a temperature of 850 °C for 1 h and subsequent cooling in water. Afterwards, the specimens were annealed (aged) for 6 h at 490 °C, thus forming precipitates. On the other hand, samples were annealed (aged) at 490 °C for 6 h without previous solution annealing.

2.5. Metallography and Microscopy

The PBF-LB/M manufactured specimens were ground parallel to the build-up direction with SiC grinding paper, from 320 to 1000 mesh. Subsequently, the ground cross-sections were polished using diamond suspensions with grain sizes from 3 to 1 µm. Finally, samples for microscopic investigations of the microstructures were etched with a V2A etching solution, a water-based mixture of HCl, $HN0_3$ and Vogel's special reagent.

A Keyence VHX 6000 light optical digital microscope from Keyence Corporation (Osaka, Japan) was used to record the cross-section of the cuboid specimens. The relative density was then quantified by phase comparison using the software ImageJ (v. 1.53j, 2021, Wayne Rasband, Kensington, MD, USA), by subtracting the accumulated pore fraction.

Microstructural investigations were performed using a type MIRA 3 field-emission scanning electron microscope from TESCAN (Brno-Kohoutovice, Czech Republic). Thereby, an acceleration voltage of U_A = 15 kV, and a working distance of WD = 15 mm was used. All micrographs were taken in secondary electron (SE) contrast.

2.6. Hardness Testing

Vickers hardness testing was performed according to DIN EN ISO 6507-1. A type KB30s of KB Prüftechnik GmbH (Hochdorf-Assenheim, Germany) was used. The normal force was set to F_N = 9.807 N (HV1). The hardness values were calculated by the mean of five indents on the samples' cross-sections, respectively.

3. Results and Discussion

3.1. Powder Feedstock Properties

The particle size distribution (PSD) of the feedstock powder exhibited a monomodal distribution with a $x_{50,3}$ of 31.8 µm, a small span of 0.94 (the span was defined as (×90,3–×10,3)/×50,3) and a nearly spherical particle shape (cf. Figure 3). However, a few satellites and potato-shaped particles are visible. In Table 4 and Figure 4, the particle size distributions of the coated powders are depicted, showing particle sizes in the same range as the feedstock material. As a result, the coating process did not affect the particle size distribution, proving that no granulation of the steel particles occurred.

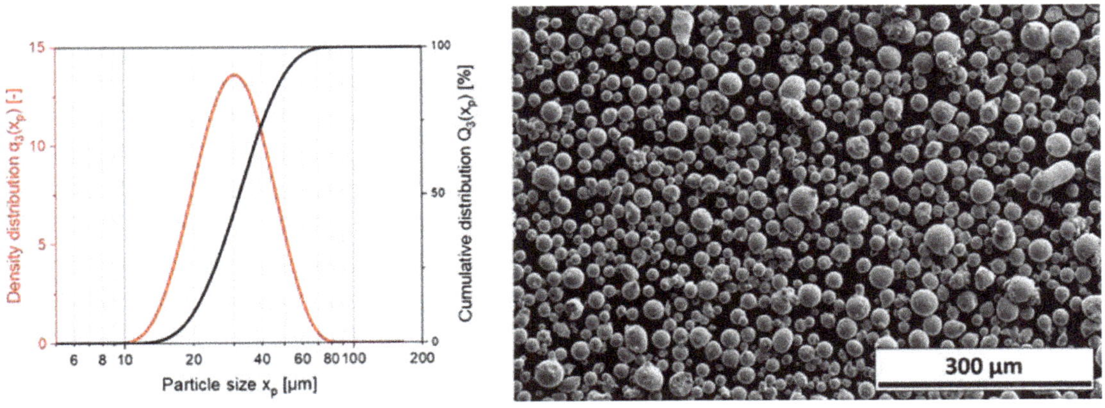

Figure 3. Feedstock powder 1.2709. Particle size distribution (**left**) SEM-picture at 500× magnification (**right**).

Table 4. Particle size distributions of feedstock powder 1.2709 and coated powders.

Powder Material	Particle Size (μm)			Span
	$x_{10,3}$	$x_{50,3}$	$x_{90,3}$	$\frac{x_{90,3}-x_{10,3}}{x_{50,3}}$ (-)
1.2709	19.9	31.8	49.7	0.94
1 vol.% SiC/1.2709	20.9	32.6	50.0	0.89
1 vol.% IOB/1.2709	20.2	32.0	49.7	0.92
0.75 vol.% FLG/1.2709	20.1	31.8	49.5	0.92

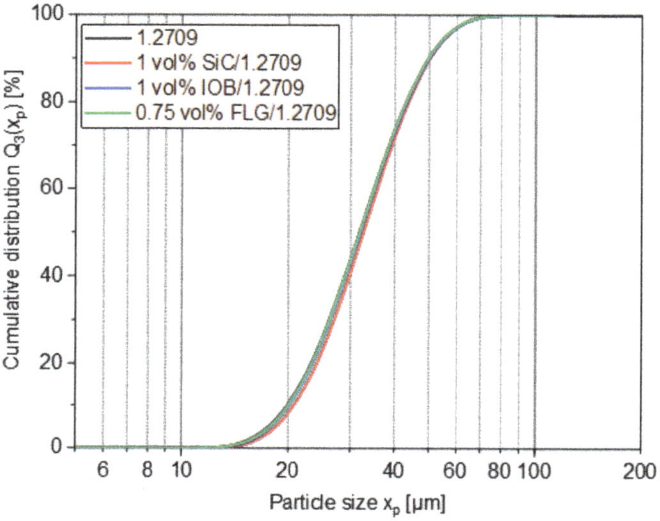

Figure 4. Particle size distribution of feedstock powder 1.2709 and nanoparticle coatings.

In Figure 5, SEM pictures of uncoated and coated particles at higher magnifications are compared. All coatings did not significantly change the particle shape. The SEM pictures also provide information about the homogeneity of the coating. In the case of SiC coating, well-distributed particles can be seen on the steel particle surfaces. For IOB, a homogenous layer enclosing the steel particle is apparent. The FLG coating exhibits a partially covered area, on which the thin FLG flakes form a closed carpet on the surface (dark appearing area on Figure 5). However, there are also visible spots without a coating by nanoparticles.

Laser Reflectance of the Feedstocks

Figure 6 shows the results of the reflectance analysis. A total of three measurement runs were performed for each of the four materials. The feedstock material 1.2709 with an average value of 10.2% exhibited the highest reflectance. The subsequent materials were represented by SiC/1.2709 with 9.2% and FLG/1.2709 with 7.9%. The material showing the lowest value of 7.2% was represented by IOB/1.2709. To ensure comparability between the measurements, a polished aluminum sample was used for the background measurement. Accordingly, a reduction in reflectance of up to 30% could be achieved. The degree of absorption of a powder material significantly depends on the PSD, morphology, surface roughness, and optical appearance (e.g., color) [26,28,35]. Due to the consistent results of the PSD, the cause is not attributable to a change in particle size. Thus, the nanoparticle coatings cause an increase in surface roughness of individual metallic powder particles. According to Karg et al. [36] and Lüddecke et al. [34] the applied nanoparticles act as spacers between the powder particles and thus, result in an increased surface roughness. This could allow easier coupling of the laser beam and enabled multiple reflectances with the powder bulk due to an increased amount of beam traps [37]. Initial measurements of the surface roughness of individual powder particles by atomic force microscopy (AFM) indicated a quantitative correlation between surface roughness and reflectance behavior, as illustrates in a publication by Lüddecke et al. [34]. In future work, this measurement will be performed on a representative number of particles (several hundred) to be able to provide a valid statement. Since the manual measurement of a section of powder particle surface requires a lot of time, a more automated solution must first be implemented.

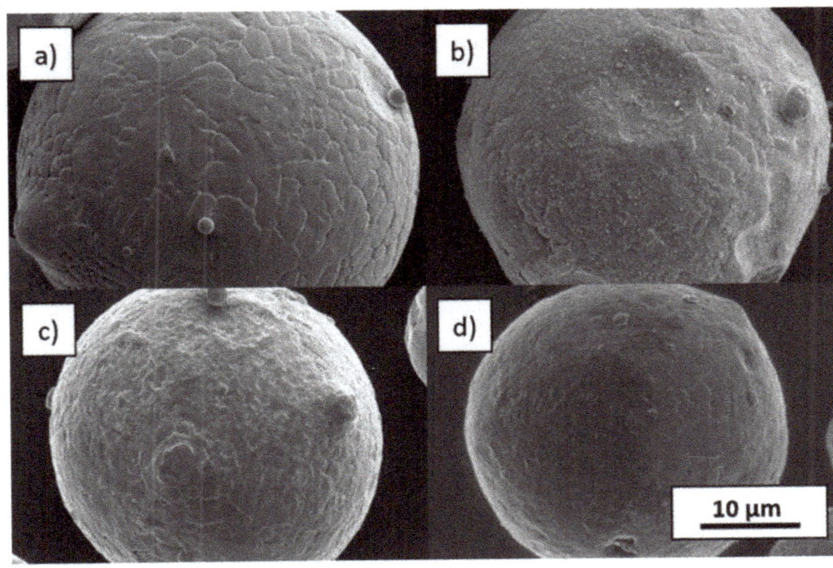

Figure 5. SEM pictures at 10,000× magnification: (**a**) feedstock 1.2709, (**b**) 1 vol.% SiC, (**c**) 1 vol.% IOB and (**d**) 0.75 vol.% FLG.

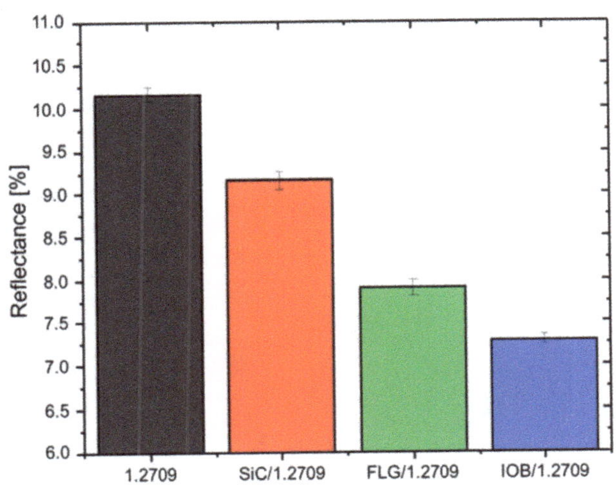

Figure 6. Reflectance measurements of the coated and uncoated tool steel 1.2709 powder at a wavenumber of 9398 cm^{-1}.

3.2. PBF-LB/M Single Track Scans

When examining the microscopic images of the individual single tracks, differences in shape, homogeneity and color can be observed. Overall, the single tracks can be classified into three categories.

- Low energy density (<3.75 J/mm^2)
- Medium energy density (3.75–15 J/mm^2)
- High energy density (>15 J/mm^2)

In Figure 7, the top view of the single tracks for three corresponding energy density levels for the 1.2709 feedstock and its nanoparticle coatings are illustrated. At an energy density of 1.5 J/mm^2, an insufficient energy input is applied to the metal powder to melt it completely. Instead, highly interrupted and irregularly shaped melt pools were observed. This behavior occurred in all material combinations discussed in this work. As a result of the poor wettability and lack of continuity of the melt pool, there was insufficient connection to the underlying layer or build platform. The side view (cf. Figure 8) shows a straight single track, and increased porosity between the layers can be observed. In Figure 9, the dimension of the melt pool width for varying energy densities is presented. In the area of an energy density <3.75 J/mm^2, an average melt pool width of 30–60 µm is measured, which corresponds to a single to double size of the focal diameter of the utilized laser. The second energy density level (5 J/mm^2) illustrated in Figures 7 and 8 shows a homogeneous and continuous melt pool for the 1.2709 feedstock, SiC/1.2709 and FLG/1.2709. However, IOB/1.2709 shows irregularities in the melt pool width. In the side view, SiC/1.2709 shows minor waviness of the top layer. Overall, all samples of this energy density level are characterized by a dense and stable single track with a solid connection between the individual layers and the build platform. The melt pool widths for all material combinations in the medium range (3.75–15 J/mm^2) are about 70–150 µm and thus 2.5–5 times the focal diameter of the laser. In the top view, the single tracks of the highest energy level (40 J/mm^2) characterized by a homogeneous but significantly increased melt pool width (cf. Figure 7). However, all single tracks showed more pronounced annealing colors due to the high energy input and its reaction with the remaining oxygen content during the PBF-LB/M process. This was particularly evident in IOB/1.2709 due to the applied oxygen-rich nanoparticle coating. Consequently, an increased oxygen content of 0.17 mass% was detectable, as depicted in Table 5.

Table 5. Chemical composition of the uncoated and coated tool steel 1.2709 feedstock measured by optical emission spectroscopy and carrier hot gas extraction (for C and O) in mass%.

	C	Ni	Co	Mo	Ti	Cr	Si	Mn	O	Fe
1.2709	0.01	17.36	9.31	4.59	1.18	0.14	0.05	0.01	0.02	67.32
SiC/1.2709	0.11	17.28	9.27	4.57	1.17	0.14	0.40	0.01	0.02	67.02
FLG/1.2709	0.15	17.34	9.30	4.58	1.18	0.14	0.05	0.01	0.02	67.23
IOB/1.2709	0.01	17.28	9.26	4.56	1.17	0.14	0.05	0.01	0.17	67.34

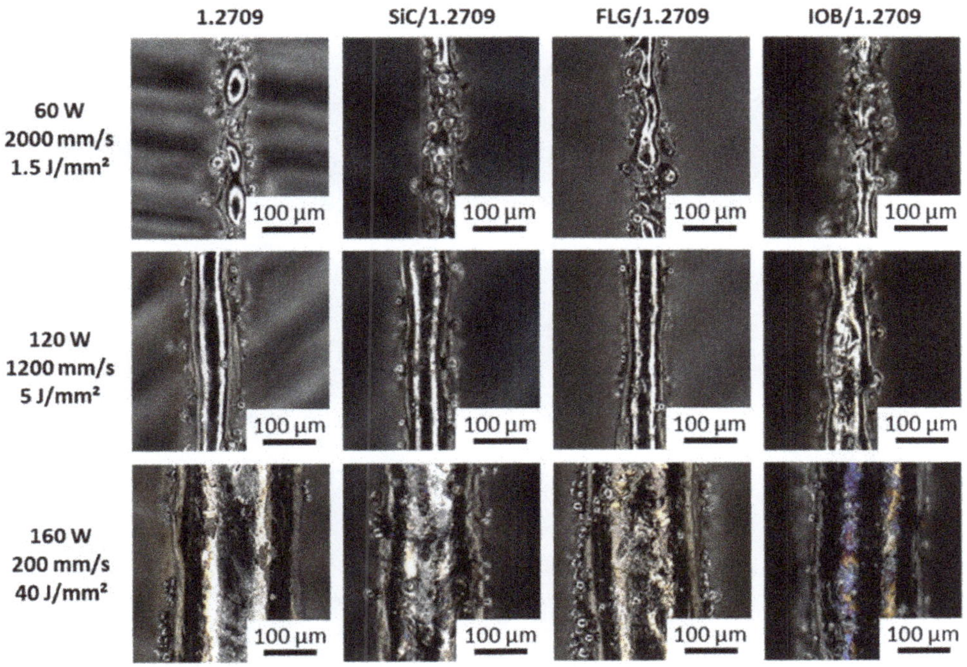

Figure 7. Top view of the single track of 1.2709, SiC/1.2709, FLG/1.2709, and IOB/1.2709.

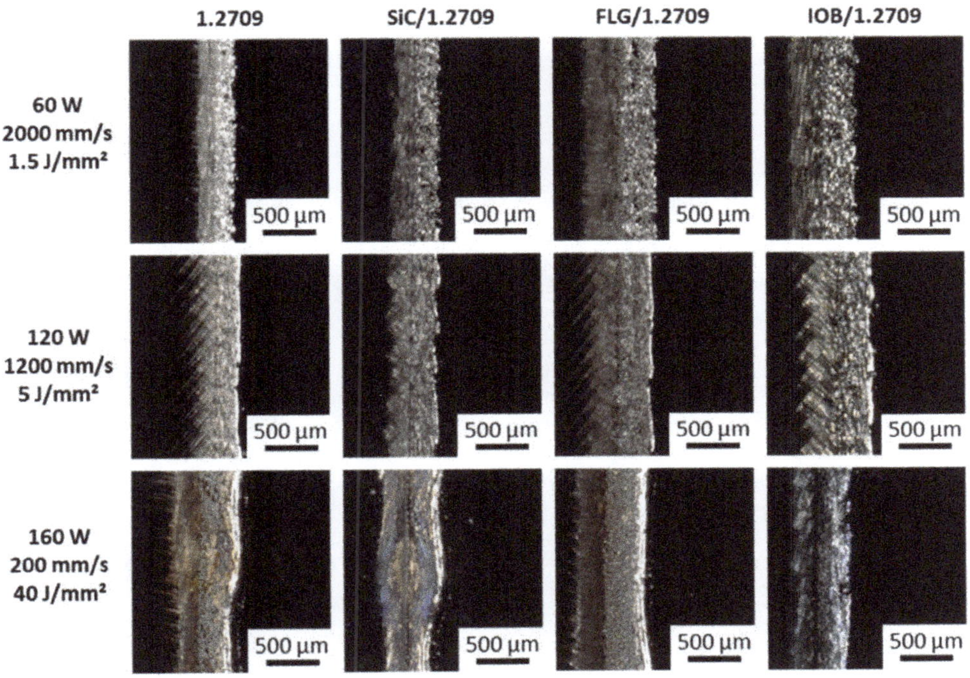

Figure 8. Side view of the single track of 1.2709, SiC/1.2709, FLG/1.2709, and IOB/1.2709.

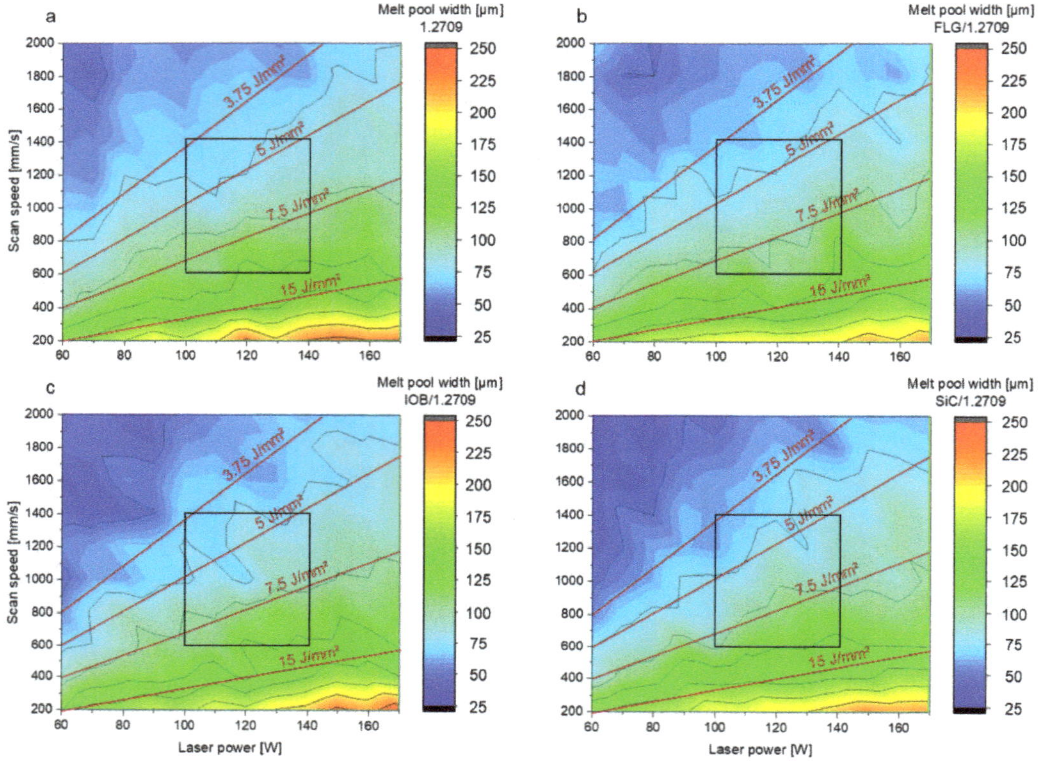

Figure 9. Melt pool width with respect to laser power and scan speed of (**a**) 1.2709 feedstock material, (**b**) FLG/1.2709 (0.75 vol.%), (**c**) IOB/1.2709 (1 vol.%), (**d**) SiC/1.2709 (1 vol.%). Parameter window for cuboid specimens highlighted by black square.

3.3. *Properties of the PBF-LB/M Densified Tool Steel*

3.3.1. PBF-LB/M Densification of the Used Tool Steel Powders

In Figure 10, the outcomes of the relative density evaluation are illustrated. All material combinations achieved a relative density of over 99.9%. The strongly reduced process window of IOB/1.2709 is conspicuous. As already shown in a previous publication [34], the coating of IOB nanoparticles exhibits pronounced hydrophilic properties. This leads to clumping of the powder particles, which significantly impairs the flowability resulting in voids during powder application. The 1.2709 feedstock, FLG/1.2709, and SiC/1.2709 exhibit a more comprehensive process window with relative densities above 99.9%. Furthermore, it is noticeable that a suitable parameter set for SiC/1.2709 appeared at lower scan speeds and with a tendency to higher laser power levels. The FLG coated 1.2709 is characterized by a broader, and more efficient, parameter window than the illustrated 1.2709 feedstock material. The generation of dense component structures is ensured by scan speeds of up to 1400 mm/s and a laser power of 120 W with a melt pool overlap of 20%. The build rate of the evaluated specimens is illustrated in Figure 11. This was calculated as the product of layer thickness, scan speed and hatch distance. First, a relative density of over 99.9% was guaranteed over a wide range of build rates. Secondly, as was already to be expected from Figure 10, an increased standard deviation of relative density of the IOB/1.2709 specimens was noticeable. A more detailed examination of the build rates over 1.5 mm^3/s showed that FLG/1.2709 allowed production with a relative density above 99.9%. With a build rate of 1.86 mm^3/s, this exceeded the 1.2709 feedstock material by approximately 18%. The SiC nanoparticle coating also achieved an increased build rate but was not represented by a reliable consistency. In Figure 12, micrographs of the cross-sectional areas of selected specimens are illustrated. Process parameters exhibiting an increased scan speed led to a more frequent lack of fusion (blue arrows). Lack of fusion defects represented insufficient melting of adjacent welding lines, which appeared after solidification in the form of primarily non-fused areas. In some cases, these may also contain unmelted powder particles [38,39]. Lack of fusion often extends within the x-y-plane or the z-direction through several layers. In the case of IOB/1.2709, this defect pattern was particularly evident. The already mentioned inhomogeneous powder distribution during the PBF-LB/M process intensified this effect. In addition, pores are noticeable in Figure 12 (red arrows). Pores usually occur in a spherical form and can be either randomly distributed in the part or appear systematically in the form of a pore seam. In this case, isolated pores were predominantly observed at a decreased velocity (1.2709 feedstock, IOB/1.2709). The low number of defects and the extended process window with FLG/1.2709 was attributable to the improved reflectance behavior and the possible increased heat conduction by the nanoparticle coating, which was already examined in more detail with an FLG/1.4404 material combination in a previous publication [26].

3.3.2. Microstructure of the PBF-LB/M Processed Tool Steel

To investigate the impact of the nanoparticle coating of the feedstock powder on the densification during PBF-LB/M in more detail, micrographs were taken using a SEM. Figure 13 shows the microstructures of the PBF-LB/M processed specimens. Between the microstructures of the 1.2709 sample and the specimens additivated by nanoparticles, no significant differences could be detected. On the mesoscale (lower magnification) fusion lines of the individual melt tracks and volumetric defects like pores (as discussed in Section 3.3.1) could be observed for all four samples. At higher magnifications, a cellular microstructure became visible (cf. Figure 13e–h). This cellular microstructure is typical for PBF-LB/M-processed steel and was also reported by Kempen et al. [40], who also investigated additively manufactured tool steel 1.2709. Adjacent cells formed colonies, which share the same crystallographic orientation [41]. The colonies were separated by large-angle grain boundaries and formed grains that could be observed at lower magnifications (Figure 13a–d).

The formation of the cellular sub-microstructure was associated with the rapid solidification rate, present in the PBF-LB/M process. The cells solidified first, followed by the surrounding seam areas (Figure 13e–h). The seams consisted of small-angle boundaries, possessed a high dislocation density and were formed as a result of the high degree of constitutional supercooling occurring during solidification [42]. After PBF-LB/M processing of steel 1.2709, Kučerová et al. observed increased contents of the elements Ni, Mo, and Ti in the seam areas [43] and Strakosova et al. showed that the cell seams are also enriched with the element C [44]. The microsegregations of the elements C, Ni, Mo, and Ti reduce the martensite start temperature locally [45]. Thus, fcc phase austenite can be stabilized at room temperature in the seam areas of the PBF-LB/M processed maraging tool steel 1.2709, which otherwise consists of bcc martensite [43]. Retained austenite is not common in conventional produced maraging steel, but was also observed in segregated inter-dendritic regions of steel 1.2709 additively manufactured using directed energy deposition (DED-LB/M) [43,46].

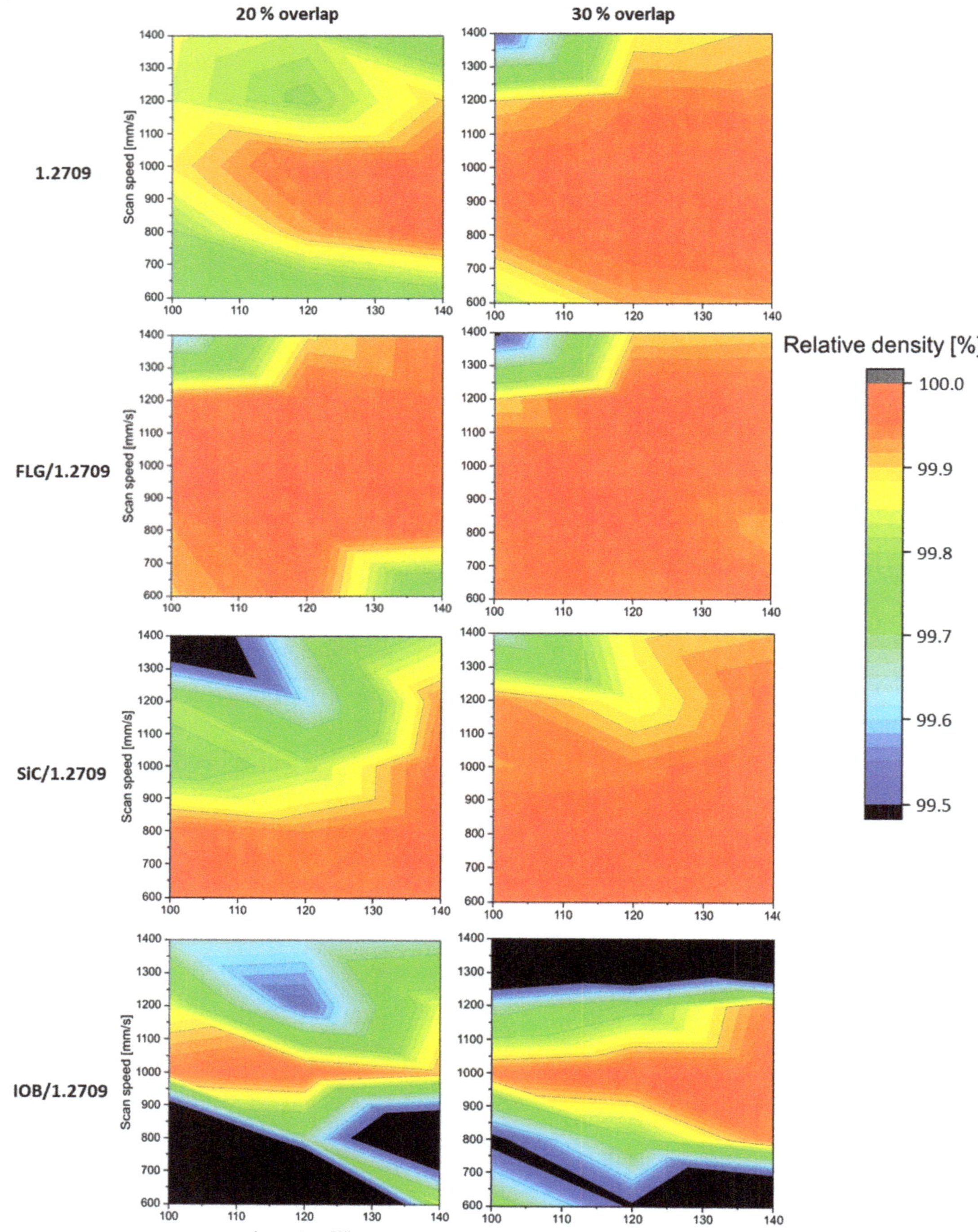

Figure 10. Relative densities of 1.2709, SiC/1.2709, FLG/1.2709, and IOB/1.2709 using different overlaps for adjacent welding lines of 20% and 30%.

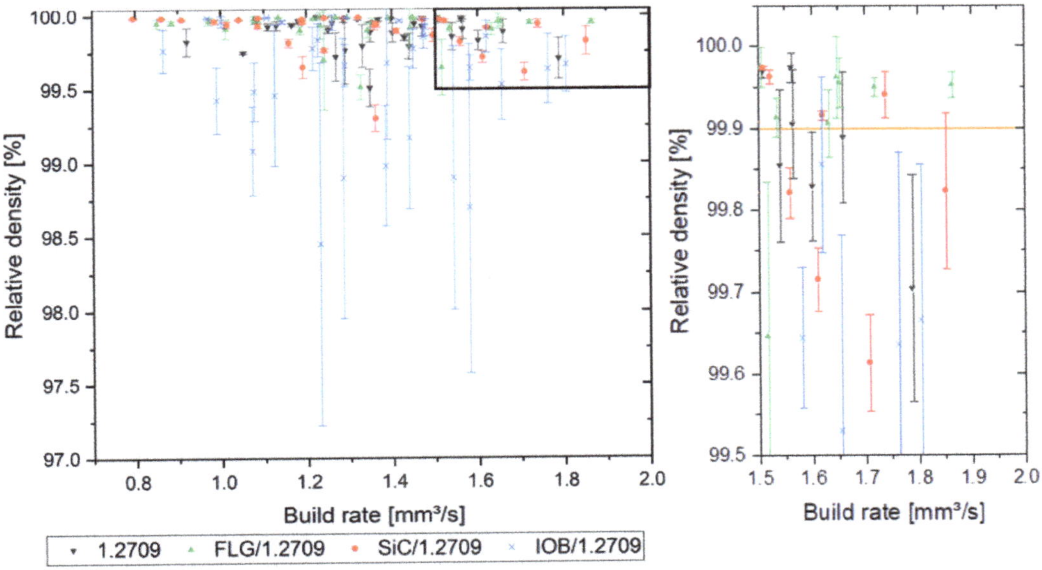

Figure 11. Relative density with respect to build rate of 1.2709 and its nanoparticle coatings. Overview of all evaluated specimens (left diagram), black framed area illustrated in right diagram.

Furthermore, C-rich segregations increase the susceptibility of the formation of carbides. Precipitations were found at the triple points of the segregated seam areas in this study. Considering the Ti and C enrichment inside the seam areas, it can be assumed that these precipitations were Ti-rich carbides of type MC. Accordingly, the precipitations seemed to be enlarged and occur more frequently for the PBF-LB/M sample made from powder FLG/1.2709. This starting powder showed an increased C content due to the C-rich FLG coating (0.15 mass% C, cf. Table 5). Like the FLG coating, the SiC coating used in this study is also rich with the element C and increased indirectly the C content of the feedstock powder, which again could promote carbide formation (0.11 mass% C, cf. Table 5). At the same time the SiC coating indirectly increased the Si content of the feedstock powder, which could counteract the carbide formation in the segregated areas of retained austenite to some extent, preventing the increase in number and size of carbide precipitations [47].

3.3.3. Microstructure and Mechanical Properties of the PBF-LB/M Processed Tool Steel in the Heat-Treated Condition

Conventionally maraging steels like the considered tool steel 1.2709 are used in solution-annealed and precipitation-hardened conditions to ensure the strength and hardness required for their applications, such as in injection molding tools for plastic processing [46]. Solution annealing is performed to generate a sufficient high solution state of alloying elements like Ti, Mo, and Ni. The high solution state is required because these elements are needed to form precipitations of intermetallic phases like Ni_3(Ti,Mo) of a specific sizes to achieve a maximum precipitation hardening effect during subsequent aging [48]. The formation of the intermetallic phases during additive manufacturing processes and subsequent heat treatments is described in detail in the work of Jägle et al. [49]. In the literature, the solution state of additively manufactured maraging steels is already high in as-built conditions because of the high solidification rate present [44]. Therefore, in this work, samples were not exclusively solution annealed and were subsequently aged for precipitation hardening, but also aged only after PBF-LB/M production. The omission of the solution annealing step can further reduce production time and yields cost savings in the additive manufacturing of tools.

Figure 14 shows the microstructures obtained after the heat-treatment of the PBF-LB/M manufactured samples. After solution annealing (850 °C, 1 h) and subsequent aging (490 °C, 6 h), the fine cellular microstructure, which is typical for PBF-LB/M processed steel, is diminished for all investigated samples, either additivated or non-additivated. Instead, a lath-like martensitic microstructure is formed. This microstructure formed corresponds with the microstructure of maraging steels, conventionally produced by casting and hot forming [46].

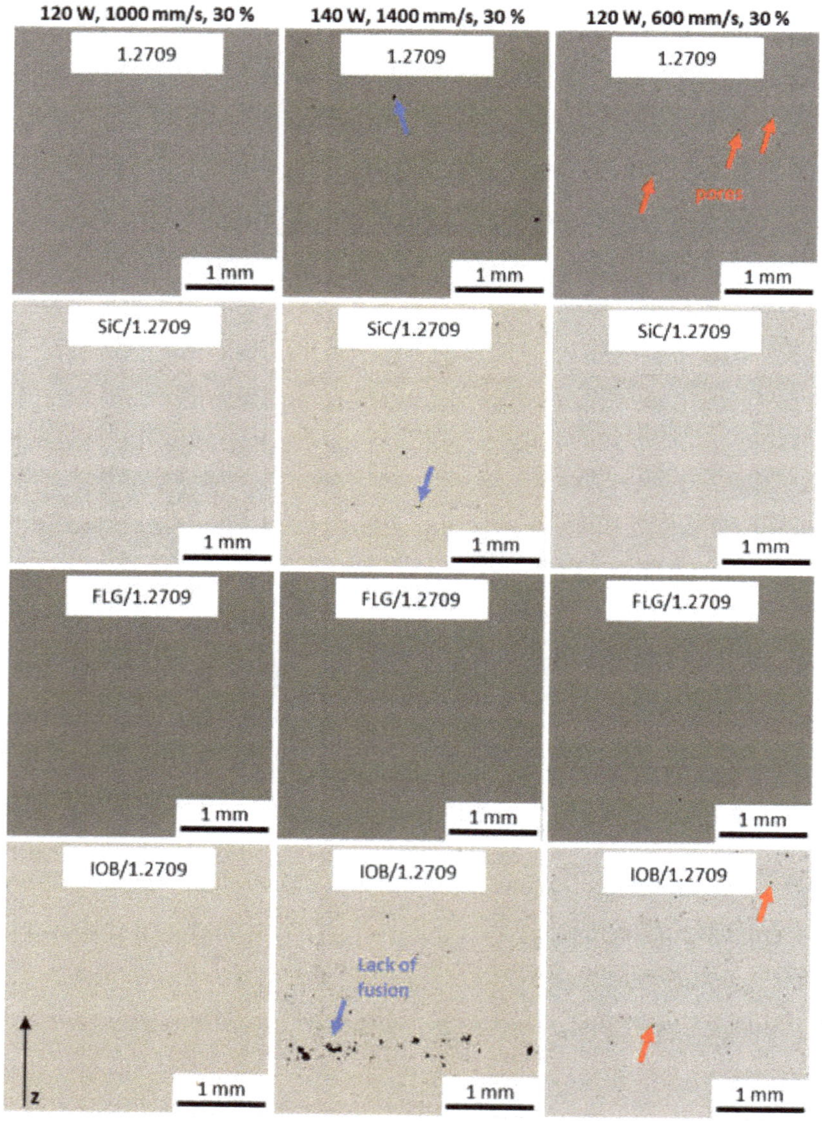

Figure 12. Cross-sections of manufactured specimens different process parameters and an overlap of 30%; pore formation indicated by red arrows; lack of fusion indicated by blue arrows.

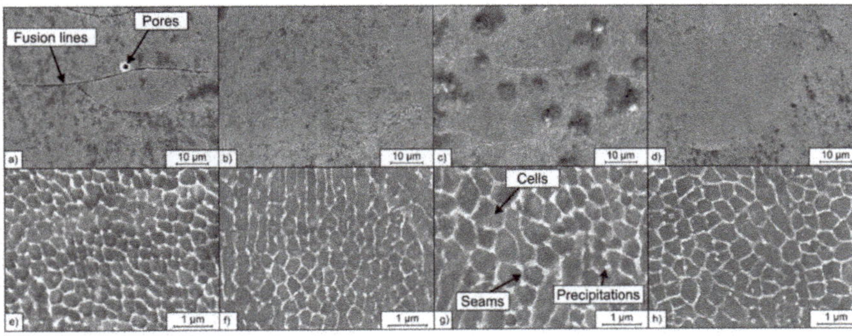

Figure 13. SEM micrographs of the PBF-LB/M densified samples at different magnifications: (**a**) 1.2709, (**b**) SiC/1.2709, (**c**) FLG/1.2709, (**d**) IOB/1.2709 at low magnification; (**e**) 1.2709, (**f**) SiC/1.2709, (**g**) FLG/1.2709, (**h**) IOB/1.2709 at a higher magnification.

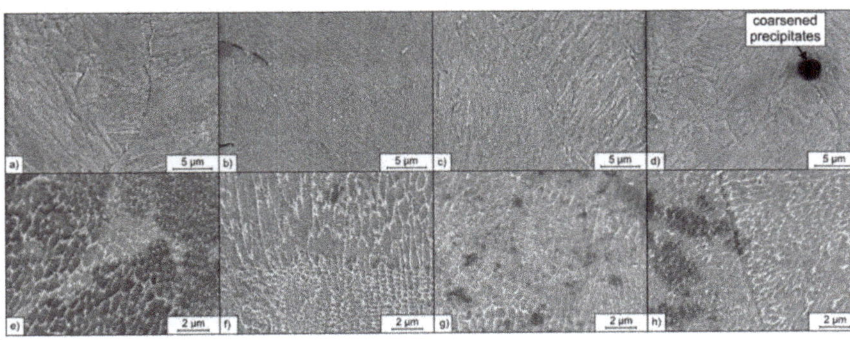

Figure 14. SEM micrographs of the PBF-LB/M densified and heat-treated samples: (**a**) 1.2709, (**b**) SiC/1.2709, (**c**) FLG/1.2709, (**d**) IOB/1.2709 solution annealed and aged; (**e**) 1.2709, (**f**) SiC/1.2709, (**g**) FLG/1.2709, (**h**) IOB/1.2709 aged only.

Differences considering the microstructures of the solution annealed and aged samples can be found for the samples made from the SiC/1.2709 and IOB/1.2709 powders. Inclusions are appearing dark in the secondary electron contrast and can be observed in the cross-sections of these samples (Figure 14b,d). It is assumed that these inclusions are oxides. Similar inclusions in PBF-LB/M-processed steel 1.2709 were identified as Ti-, Mo-, Al-, and Si-rich oxides by Kempen et al. [40]. For the sample made from the powder SiC/1.2709, the oxide precipitations are of a stretched-out shape with a length of approx. 3 µm. In contrast, the oxides in the sample made from IOB/1.2709 are of a spherical morphology with diameters up to 5 µm. The occurrence of these enlarged oxides can be traced back to the increased oxygen content of the IOB coated feedstock powder IOB/1.2709 (0.17 mass% O, cf. Table 5). The nanoparticle coating seems to be dissolved during the PBF-LB/M process, resulting in oxygen enrichment of the melt. The high oxygen content subsequently promotes the formation and coarsening of oxides during the solution annealing. Consequently, the enlarged oxides could not be observed for the samples made from SiC/1.2709 and IOB/1.2709 in the as-built and aged condition (Figure 14e–h). The pronounced tempering colors observed for the samples made from IOB/1.2709 during the single-track experiments (Section 3.2) indicate that a certain number of O-rich particles is carried to the surface of a selectively melted layer by convection in the melt pool. This can be traced back to the lower density of the incompletely dissolved oxide particles (IOB) compared to the molten alloy. The primary assumption is that an insufficient powder deposition in this particular case is associated with volumetric defects like pores and lacks of fusion, and not with the occurrence of oxides.

In contrast to the microstructures obtained after solution annealing and precipitation hardening, the fine cellular microstructure previously formed during the PBF-LB/M densification is still present after aging (490 °C, 6 h) for all investigated samples (Figure 14e–h). This behavior was also observed by Strakosova et al. [44], who explained the preservation or diminishing of the cellular microstructure by diffusion processes of the alloying elements depending on the heat-treatment temperature. According to Strakosova et al. [44], the segregated cell seams become homogenized due to the possibility of diffusing of the involved elements (Ni, Mo, Ti, C) during the solution annealing temperatures around 850 °C. But the relatively lower aging temperature of around 490 °C is not sufficient for homogenization of the segregated areas by diffusion processes. Therefore, aging alone does not diminish the cellular microstructure [44].

While a fine microstructure like the cellular microstructure can increase the material's strength and toughness, the inhomogeneous chemical element distribution after aging can result in a reduced achievable hardness due to the stabilizing of retained austenite [46,50]. Strakosova et al. [44] observed an increased amount of retained austenite only after aging compared to a solution annealed and aged state owing to the segregated seam areas which are still present after aging only. However, since the formation of carbon-martensite is not the dominant hardening mechanism in soft martensitic maraging steels due to their low C content, the presence of the retained austenite does not reduce the hardness of only-aged samples significantly compared to samples, which were homogenized during a prior solution annealing. The high hardness of maraging steels is achieved by the precipitation hardening during final aging [44].

The results of Strakosova et al. [44] could be confirmed in this work, as Figure 15 shows. The hardness of the PBF-LB/M processed steel 1.2709 drops only by 4 HV1 from 666 HV1 in the solution annealed and aged condition to 662 HV1 in the aged-only condition. However, the samples produced from the SiC/1.2709 and IOB/1.2709 powders show a higher drop in hardness for the aged-only condition compared to the solution annealed and aged state from 672 to 630 HV1, respectively from 650 to 630 HV1. In contrast, the samples made from the FLG/1.2709 powder show an almost identical hardness in both heat-treated conditions solution annealed plus aged and aged-only of 655 HV1 and 654 HV1, respectively.

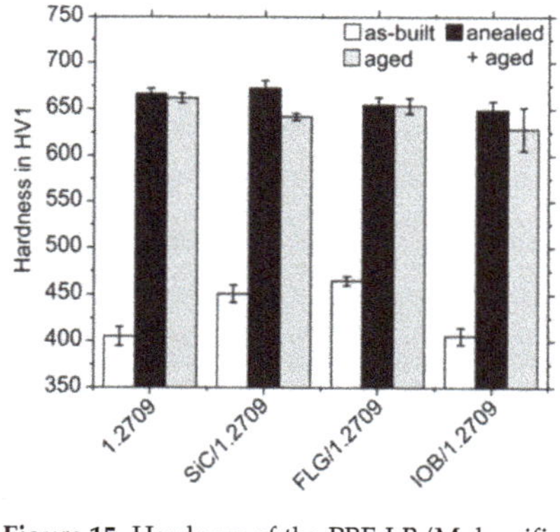

Figure 15. Hardness of the PBF-LB/M densified samples in different conditions.

The hardness in the as-built state of the PBF-LB/M-processed samples was also determined. While the hardness values of the samples made from standard 1.2709 and the IOB/1.2709 powder are identical with 405 HV1, the hardness values of the SiC and FLG additivated samples in as-built condition are higher with values of 451 HV1 and 465 HV1, respectively. The increased hardness of the samples made from the SiC and FLG coated powders can be traced back to the increased content of the element C compared to the standard 1.2709 powder (cf. Table 5). The element C is an interstitial soluble element and causes strong solid solution hardening in the as-built condition of the samples SiC/1.2709 and FLG/1.2709. But the increased hardness due to solid solution hardening respectively carbon-martensitic hardening caused by the increased C contents does not improve the hardness in the heat-treated condition. A reason for this is that the predominant hardening mechanism of heat-treated maraging steels is not a solid-solution hardening by intestinally solute C or a car-bon-martensitic hardening. Instead, the hardening of maraging steels relies on the mechanism of precipitation hardening by the precipitation of intermetallic phases like Ni_3(Ti, Mo) of a specific size during the aging heat-treatment [49].

Furthermore, the aforementioned increased susceptibility for the formation of Ti-rich carbides owing to the increased C content of the powders coated with C-rich nanoparticles could promote a less pronounced precipitation hardening effect during aging. It could be assumed that an increased fraction of thermodynamically stable Ti-rich MC carbides reduces the amount of solute Ti in the metal matrix that is available for the formation of the intermetallic phase Ni_3(Ti, Mo), which is responsible for the precipitation hardening of steel 1.2709 [51]. However, in this study, no significant hardness decrease in the heat-treated condition could be observed using the C-rich nanoparticle additive FLG. This invalidates the assumption of a hardness loss owing to the pronounced formation of Ti-rich MC carbides.

The IOB coating can also result in a decreased hardness owing to the aforementioned negative impact on the flowability leading to a less dense powder application during PBF-LB/M processing [34]. Therefore, the IOB coating can promote lack of fusion defects which can affect the Vickers hardness testing, resulting in a decreased measured hardness, as reported by Kempen et al. [40]. Indeed, the hardness decreasing effect of volumetric defects could be observed for the sample made from IOB/1.2709 in the aged-only condition (cf. Figure 15). The local distribution of the lack of fusion defects is the reason for the high scattering of the hardness values (standard deviation of 23 HV1). Simultaneously, other sample areas can possess a higher density with less lack of fusion defects and thus show no decrease in hardness and a lower scattering (cf. Figure 15).

The FLG nanoparticle coating of 1.2709 powder enables more efficient PBF-LB/M process parameters which increase the build rate by 18% while preserving the material's high hardness after precipitation hardening. Thereby, the solution annealing, which is performed in the conventional production of maraging steels, can be omitted. In contrast, nanoparticle coating of the feedstock material using IOB seems inappropriate because of the promotion of oxide formation and lack of fusion defects. Such defects can have a detrimental impact on tensile and fatigue properties. In future studies, the evolution of the microstructure and the formation of the precipitations has to be investigated more deeply by techniques like transmission electron microscopy (dislocations, precipitations) and electron backscatter diffraction (retained austenite, martensite, precipitations) in order to improve the heat-treatment and thus the final part properties. Additionally, mechanical testing, including tensile and fatigue tests, will be performed in future work.

4. Conclusions

In this work, nanoparticle (SiC, FLG, IOB) coated tool steel 1.2709 powder was analyzed concerning its particle size distribution and reflectance behavior and was processed by PBF-LB/M. The subsequent examination of the relative density, build rate, hardness, and microstructure under consideration of different heat-treatment methodologies allows the determination of production efficiency. Consequently, the scientific questions addressed at the beginning are answered:

1. After nanoparticle coating of the metallic powder particles, increased absorption behavior is observed. One the one hand, the nanoparticles on the surface of individual metal powder particles lead to increased surface roughness. This in turn leads to increased beam traps and multiple scattering of laser radiation within the powder bed. On the other hand, there is a correlation between the resulting darker coloration of the powder particles and the reduced reflection at the utilized wavelength of 1064 nm as an additional attribute.

2. Based on DRIFTS analysis, IOB/1.2709 exhibits the lowest reflectance values. However, the relative density analysis of PBF-LB/M produced samples reveals the smallest process window for this composite powder. In addition to the reflectance, a homogeneous powder bed is of great importance. In this case, its inferior flowability led to voids during powder application. The coating with FLG enables build rates allowing a relative density of over 99.9%, which exceeds those of the original feedstock by approximately 18%. The combination of low reflectance and increased thermal conductivity represents favorable conditions for the PBF-LB/M process. Thus, the improvement of the absorption behavior cannot be used as the sole factor to qualify more efficient process parameters.
3. A relative density of 99.9% was achieved with all material combinations. The generation of single tracks, which are exposed over 25 layers, represents a process-oriented and transferable qualification methodology due to the consideration of heat balance, real layer thickness and consistent diffusion process without the influence of the build platform material.
4. The microstructure of the all specimens shows a cellular substructure in the as-built condition. Furthermore, presented precipitations seem to be enlarged and occur more frequently for the PBF-LB/M sample made from powder FLG/1.2709 due to a C enrichment inside the seam areas. After solution annealing and subsequent aging, the fine cellular microstructure which is typical for PBF-LB/M processed steel, is diminished for all investigated samples, either additivated or non-additivated. Nevertheless, the IOB coating tends to promote the formation of oxides. Considering the hardness testing, FLG/1.2709 maintains the hardness of the additively manufactured and heat-treated 1.2709 feedstock material. The conventionally performed solution annealing could be omitted.

5. Outlook

Future work will investigate the correlation between surface roughness of individual powder particles and reflectance in more detail. The use and adaption of the AFM measurement to a statistically representative number of particles represent a significant factor. Furthermore, the material tests will be extended to include tensile and fatigue tests. In addition, microstructural characteristics have to be investigated more deeply by techniques like transmission electron microscopy and electron backscatter diffraction.

Author Contributions: Project administration, J.T.S. and A.K.; conceptualization, O.P. and F.G.; methodology, O.P., F.G. and A.L.; investigation, O.P., F.G. and A.L.; resources, J.T.S., A.R., F.G. and A.K.; writing—original draft preparation, O.P., F.G. and A.L.; writing—review and editing, O.P., F.G., A.L., A.R. and J.T.S.; supervision, J.T.S. and A.K.; funding acquisition, J.T.S. and A.K. All authors have read and agreed to the published version of the manuscript.

Funding: This work was supported by the Deutsche Forschungsgemeinschaft (DFG, German Research Foundation), priority program "Materials for Additive Manufacturing", SPP 2122, Grant-ID SE 2935/1-1, Project-ID 410107213. The authors gratefully acknowledge the financial supports. In addition, we acknowledge support by the Open Access Publication Funds of the Ruhr-Universität Bochum.

Institutional Review Board Statement: Not applicable.

Informed Consent Statement: Not applicable.

Data Availability Statement: The data presented in this study are available on request from the corresponding author. The data are not publicly available due to privacy.

Conflicts of Interest: The authors declare no conflict of interest.

References

1. Deutsches Institut für Normung e. V. *Additive manufacturing—General Principles—Terminology*; Beuth Verlag GmbH: Berlin, Germany, 2018.
2. Gibson, I.; Rosen, D.; Stucker, B. *Additive Manufacturing Technologies: 3D Printing, Rapid Prototyping and Direct Digital Manufacturing*, 2nd ed.; Springer: New York, NY, USA, 2015.
3. Kusoglu, I.M.; Gökce, B.; Barcikowski, S. Use of (nano-)additives in Laser Powder Bed Fusion of Al powder feedstocks: Research directions within the last decade. *Procedia CIRP* **2020**, *94*, 11–16. [CrossRef]
4. Gu, D.; Meng, G.; Li, C.; Meiners, W.; Poprawe, R. Selective laser melting of TiC/Ti bulk nanocomposites: Influence of nanoscale reinforcement. *Scr. Mater.* **2012**, *67*, 185–188. [CrossRef]
5. Gu, D.; Wang, H.; Chang, F.; Dai, D.; Yuan, P.; Hagedorn, Y.-C.; Meiners, W. Selective Laser Melting Additive Manufacturing of TiC/AlSi10Mg Bulk-form Nanocomposites with Tailored Microstructures and Properties. *Phys. Procedia* **2014**, *56*, 108–116. [CrossRef]
6. Gu, D.; Wang, H.; Zhang, G. Selective Laser Melting Additive Manufacturing of Ti-Based Nanocomposites: The Role of Nanopowder. *Met. Mater. Trans. A* **2013**, *45*, 464–476. [CrossRef]
7. AlMangour, B.; Grzesiak, D.; Yang, J.-M. Selective laser melting of TiC reinforced 316L stainless steel matrix nanocomposites: Influence of starting TiC particle size and volume content. *Mater. Des.* **2016**, *104*, 141–151. [CrossRef]
8. AlMangour, B.; Grzesiak, D.; Cheng, J.; Ertas, Y. Thermal behavior of the molten pool, microstructural evolution, and tribological performance during selective laser melting of TiC/316L stainless steel nanocomposites: Experimental and simulation methods. *J. Mater. Process. Technol.* **2018**, *257*, 288–301. [CrossRef]
9. AlMangour, B.; Grzesiak, D.; Yang, J.-M. Rapid fabrication of bulk-form TiB 2/316L stainless steel nanocomposites with novel reinforcement architecture and improved performance by selective laser melting. *J. Alloys Compd.* **2016**, *680*, 480–493. [CrossRef]
10. Zhao, Z.; Li, J.; Bai, P.; Qu, H.; Liang, M.; Liao, H.; Wu, L.; Huo, P.; Liu, H.; Zhang, J. Microstructure and Mechanical Properties of TiC-Reinforced 316L Stainless Steel Composites Fabricated Using Selective Laser Melting. *Metals* **2019**, *9*, 267. [CrossRef]
11. Doñate-Buendia, C.; Streubel, R.; Kürnsteiner, P.; Wilms, M.B.; Stern, F.; Tenkamp, J.; Bruder, E.; Barcikowski, S.; Gault, B.; Durst, K.; et al. Effect of nanoparticle additivation on the microstructure and microhardness of oxide dispersion strengthened steels produced by laser powder bed fusion and directed energy deposition. *Procedia CIRP* **2020**, *94*, 41–45. [CrossRef]
12. Chang, F.; Gu, D.; Dai, D.; Yuan, P. Selective laser melting of in-situ Al4SiC4 + SiC hybrid reinforced Al matrix composites: Influence of starting SiC particle size. *Surf. Coat. Technol.* **2015**, *272*, 15–24. [CrossRef]
13. Sehrt, J.T.; Kleszczynski, S.; Notthoff, C. Nanoparticle improved metal materials for additive manufacturing. *Prog. Addit. Manuf.* **2017**, *2*, 179–191. [CrossRef]
14. Sehrt, J.T.; Kleszczynski, S.; Notthoff, C.; Lau, M.; Gökce, B.; Barcikowski, S. Laser powder bed fusion of nano-WC-modified and nano-TiO2-modified metal powders. *Proc. Int. Conf. Prog. Addit. Manuf.* **2016**, *6*, 26–38.
15. Jadhav, S.D.; Dadbakhsh, S.; Vleugels, J.; Hofkens, J.; Van Puyvelde, P.; Yang, S.; Kruth, J.-P.; Van Humbeeck, J.; Vanmeensel, K. Influence of Carbon Nanoparticle Addition (and Impurities) on Selective Laser Melting of Pure Copper. *Materials* **2019**, *12*, 2469. [CrossRef]
16. Jadhav, S.D.; Dhekne, P.P.; Dadbakhsh, S.; Kruth, J.-P.; Van Humbeeck, J.; Vanmeensel, K. Surface Modified Copper Alloy Powder for Reliable Laser-based Additive Manufacturing. *Addit. Manuf.* **2020**, *35*, 101418. [CrossRef]
17. Jadhav, S.D.; Dadbakhsh, S.; Chen, R.; Shabadi, R.; Kruth, J.-P.; Van Humbeeck, J.; Vanmeensel, K. Modification of Electrical and Mechanical Properties of Selective Laser-Melted CuCr0.3 Alloy Using Carbon Nanoparticles. *Adv. Eng. Mater.* **2020**, *22*, 1900946. [CrossRef]
18. Zhang, B.; Dembinski, L.; Coddet, C. The study of the laser parameters and environment variables effect on mechanical properties of high compact parts elaborated by selective laser melting 316L powder. *Mater. Sci. Eng. A* **2013**, *584*, 21–31. [CrossRef]
19. Sehrt, J.T. Möglichkeiten und Grenzen bei der generativen Herstellung metallischer Bauteile durch das Strahlschmelzverfahren. Ph.D. Thesis, University of Duisburg-Essen, Aachen, Germany, 2010.
20. Fu, G.; Zhang, D.Z.; He, A.N.; Mao, Z.; Zhang, K. Finite Element Analysis of Interaction of Laser Beam with Material in Laser Metal Powder Bed Fusion Process. *Materials* **2018**, *11*, 765. [CrossRef]
21. Tran, T.Q.; Chinnappan, A.; Lee, J.K.Y.; Loc, N.H.; Tran, L.T.; Wang, G.; Kumar, V.V.; Jayathilaka, W.A.D.M.; Ji, D.; Doddamani, M.; et al. 3D Printing of Highly Pure Copper. *Metals* **2019**, *9*, 756. [CrossRef]
22. Prasad, H.S.; Brueckner, F.; Volpp, J.; Kaplan, A.F.H. Laser metal deposition of copper on diverse metals using green laser sources. *Int. J. Adv. Manuf. Technol.* **2020**, *107*, 1559–1568. [CrossRef]
23. Yang, Y.; Gu, D.; Dai, D.; Ma, C. Laser energy absorption behavior of powder particles using ray tracing method during selective laser melting additive manufacturing of aluminum alloy. *Mater. Des.* **2018**, *143*, 12–19. [CrossRef]
24. Boley, C.D.; Khairallah, S.A.; Rubenchik, A.M. Calculation of laser absorption by metal powders in additive manufacturing. *Appl. Opt.* **2015**, *54*, 2477–2482. [CrossRef]
25. Gu, D.; Yang, Y.; Xi, L.; Yang, J.; Xia, M. Laser absorption behavior of randomly packed powder-bed during selective laser melting of SiC and TiB2 reinforced Al matrix composites. *Opt. Laser Technol.* **2019**, *119*, 105600. [CrossRef]
26. Pannitz, O.; Lüddecke, A.; Kwade, A.; Sehrt, J.T. Investigation of the in situ thermal conductivity and absorption behavior of nanocomposite powder materials in laser powder bed fusion processes. *Mater. Des.* **2021**, *201*, 109530. [CrossRef]

27. Zhou, W.; Sun, X.; Kikuchi, K.; Nomura, N.; Yoshimi, K.; Kawasaki, A. Carbon nanotubes as a unique agent to fabricate nanoceramic/metal composite powders for additive manufacturing. *Mater. Des.* **2018**, *137*, 276–285. [CrossRef]
28. Gruber, K.; Smolina, I.; Kasprowicz, M.; Kurzynowski, T. Evaluation of Inconel 718 Metallic Powder to Optimize the Reuse of Powder and to Improve the Performance and Sustainability of the Laser Powder Bed Fusion (LPBF) Process. *Materials* **2021**, *14*, 1538. [CrossRef]
29. Di, W.; Yongqiang, Y.; Xubin, S.; Yonghua, C. Study on energy input and its influences on single-track, multi-track, and multi-layer in SLM. *Int. J. Adv. Manuf. Technol.* **2012**, *58*, 1189–1199. [CrossRef]
30. Wei, P.; Wei, Z.; Chen, Z.; Du, J.; He, Y.; Li, J.; Zhou, Y. The AlSi10Mg samples produced by selective laser melting: Single track, densification, microstructure and mechanical behavior. *Appl. Surf. Sci.* **2017**, *408*, 38–50. [CrossRef]
31. Shi, X.; Ma, S.; Liu, C.; Wu, Q. Parameter optimization for Ti-47Al-2Cr-2Nb in selective laser melting based on geometric characteristics of single scan tracks. *Opt. Laser Technol.* **2017**, *90*, 71–79. [CrossRef]
32. Kempen, K.; Thijs, L.; Van Humbeeck, J.; Kruth, J.-P. Processing AlSi10Mg by selective laser melting: Parameter optimisation and material characterisation. *Mater. Sci. Technol.* **2015**, *31*, 917–923. [CrossRef]
33. Dilip, J.J.S.; Zhang, S.; Teng, C.; Zeng, K.; Robinson, C.; Pal, D.; Stucker, B. Influence of processing parameters on the evolution of melt pool, porosity, and microstructures in Ti-6Al-4V alloy parts fabricated by selective laser melting. *Prog. Addit. Manuf.* **2017**, *2*, 157–167. [CrossRef]
34. Lüddecke, A.; Pannitz, O.; Zetzener, H.; Sehrt, J.T.; Kwade, A. Powder properties and flowability measurements of tailored nanocomposites for powder bed fusion applications. *Mater. Des.* **2021**, *202*, 109536. [CrossRef]
35. Zhang, J.; Gu, D.; Yang, Y.; Zhang, H.; Chen, H.; Dai, D.; Lin, K. Influence of Particle Size on Laser Absorption and Scanning Track Formation Mechanisms of Pure Tungsten Powder During Selective Laser Melting. *Engineering* **2019**, *5*, 736–745. [CrossRef]
36. Karg, M.C.H.; Munk, A.; Ahuja, B.; Backer, M.V.; Schmitt, J.P.; Stengel, C.; Kuryntsev, S.V.; Schmidt, M. Expanding particle size distribution and morphology of aluminium-silicon powders for Laser Beam Melting by dry coating with silica nanoparticles. *J. Mater. Process. Technol.* **2019**, *264*, 155–171. [CrossRef]
37. Kruth, J.P.; Wang, X.; Laoui, T.; Froyen, L. Lasers and materials in selective laser sintering. *Assem. Autom.* **2003**, *23*, 357–371. [CrossRef]
38. Liu, Q.C.; Elambasseril, J.; Sun, S.J.; Leary, M.; Brandt, M.; Sharp, P.K. The Effect of Manufacturing Defects on the Fatigue Behaviour of Ti-6Al-4V Specimens Fabricated Using Selective Laser Melting. *Adv. Mater. Res.* **2014**, *891-892*, 1519–1524. [CrossRef]
39. Mukherjee, T.; Zuback, J.S.; De, A.; Debroy, T. Printability of alloys for additive manufacturing. *Sci. Rep.* **2016**, *6*, 19717. [CrossRef]
40. Kempen, K.; Yasa, E.; Thijs, L.; Kruth, J.-P.; Van Humbeeck, J. Microstructure and mechanical properties of Selective Laser Melted 18Ni-300 steel. *Phys. Procedia* **2011**, *12*, 255–263. [CrossRef]
41. Krakhmalev, P.; Yadroitsava, I.; Fredriksson, G.; Yadroitsev, I. In situ heat treatment in selective laser melted martensitic AISI 420 stainless steels. *Mater. Des.* **2015**, *87*, 380–385. [CrossRef]
42. Krakhmalev, P.; Fredriksson, G.; Svensson, K.; Yadroitsev, I.; Yadroitsava, I.; Thuvander, M.; Peng, R. Microstructure, Solidification Texture, and Thermal Stability of 316 L Stainless Steel Manufactured by Laser Powder Bed Fusion. *Metals* **2018**, *8*, 643. [CrossRef]
43. Kučerová, L.; Zetková, I.; Jandová, A.; Bystrianský, M. Microstructural characterisation and in-situ straining of additive-manufactured X3NiCoMoTi 18-9-5 maraging steel. *Mater. Sci. Eng. A* **2019**, *750*, 70–80. [CrossRef]
44. Strakosova, A.; Kubásek, J.; Michalcová, A.; Průša, F.; Vojtěch, D.; Dvorský, D. High Strength X3NiCoMoTi 18-9-5 Maraging Steel Prepared by Selective Laser Melting from Atomized Powder. *Materials* **2019**, *12*, 4174. [CrossRef]
45. Barbier, D. Extension of the Martensite Transformation Temperature Relation to Larger Alloying Elements and Contents. *Adv. Eng. Mater.* **2014**, *16*, 122–127. [CrossRef]
46. Jägle, E.A.; Sheng, Z.; Kürnsteiner, P.; Ocylok, S.; Weisheit, A.; Raabe, D. Comparison of Maraging Steel Micro- and Nanostructure Produced Conventionally and by Laser Additive Manufacturing. *Materials* **2016**, *10*, 8. [CrossRef]
47. Li, Y.; Chen, X. Microstructure and mechanical properties of austempered high silicon cast steel. *Mater. Sci. Eng. A* **2001**, *308*, 277–282. [CrossRef]
48. Filho, V.X.L.; Barros, I.F.; De Abreu, H.F.G. Influence of Solution Annealing on Microstructure and Mechanical Properties of Maraging 300 Steel. *Mater. Res.* **2016**, *20*, 10–14. [CrossRef]
49. Jägle, E.A.; Choi, P.-P.; Van Humbeeck, J.; Raabe, D. Precipitation and austenite reversion behavior of a maraging steel produced by selective laser melting. *J. Mater. Res.* **2014**, *29*, 2072–2079. [CrossRef]
50. Hansen, N. Hall–Petch relation and boundary strengthening. *Scr. Mater.* **2004**, *51*, 801–806. [CrossRef]
51. Jägle, E.A.; Sheng, Z.; Wu, L.; Lu, L.; Risse, J.; Weisheit, A.; Raabe, D. Precipitation Reactions in Age-Hardenable Alloys During Laser Additive Manufacturing. *JOM* **2016**, *68*, 943–949. [CrossRef]

materials

Article

Development of a Laser Powder Bed Fusion Process Tailored for the Additive Manufacturing of High-Quality Components Made of the Commercial Magnesium Alloy WE43

Stefan Julmi [1,*], Arvid Abel [2], Niklas Gerdes [2], Christian Hoff [2], Jörg Hermsdorf [2], Ludger Overmeyer [2], Christian Klose [1] and Hans Jürgen Maier [1]

1 Institut für Werkstoffkunde (Materials Science), Leibniz University Hanover, 30823 Garbsen, Germany; klose@iw.uni-hannover.de (C.K.); maier@iw.uni-hannover.de (H.J.M.)

2 Laser Zentrum Hannover e.V., 30419 Hannover, Germany; a.abel@lzh.de (A.A.); n.gerdes@lzh.de (N.G.); c.hoff@lzh.de (C.H.); j.hermsdorf@lzh.de (J.H.); l.overmeyer@lzh.de (L.O.)

* Correspondence: julmi@iw.uni-hannover.de; Tel.: +49-511-762-4328

Abstract: Additive manufacturing (AM) has become increasingly important over the last decade and the quality of the products generated with AM technology has strongly improved. The most common metals that are processed by AM techniques are steel, titanium (Ti) or aluminum (Al) alloys. However, the proportion of magnesium (Mg) in AM is still negligible, possibly due to the poor processability of Mg in comparison to other metals. Mg parts are usually produced by various casting processes and the experiences in additive manufacturing of Mg are still limited. To address this issue, a parameter screening was conducted in the present study with experiments designed to find the most influential process parameters. In a second step, these parameters were optimized in order to fabricate parts with the highest relative density. This experiment led to processing parameters with which specimens with relative densities above 99.9% could be created. These high-density specimens were then utilized in the fabrication of test pieces with several different geometries, in order to compare the material properties resulting from both the casting process and the powder bed fusion (PBF-LB) process. In this comparison, the compositions of the occurring phases and the alloys' microstructures as well as the mechanical properties were investigated. Typically, the microstructure of metal parts, produced by PBF-LB, consisted of much finer grains compared to as-cast parts. Consequently, the strength of Mg parts generated by PBF-LB could be further increased.

Keywords: powder bed fusion; magnesium; process development

Citation: Julmi, S.; Abel, A.; Gerdes, N.; Hoff, C.; Hermsdorf, J.; Overmeyer, L.; Klose, C.; Maier, H.J. Development of a Laser Powder Bed Fusion Process Tailored for the Additive Manufacturing of High-Quality Components Made of the Commercial Magnesium Alloy WE43. *Materials* **2021**, *14*, 887. https://doi.org/10.3390/ma14040887

Academic Editors: Bilal Gökce and Jai-Sung Lee

Received: 8 January 2021
Accepted: 8 February 2021
Published: 13 February 2021

1. Introduction

The role of metals in additive manufacturing (AM) technology has increased in recent years [1]. The production of metal products by AM is divided into three methods, i.e., powder bed systems, powder feed systems and wire feed systems. The advantages of powder and wire feed systems are the possibility to easily repair parts and having higher volume build-up rates. However, the most common method is the powder bed system or powder bed fusion (PBF-LB) according to DIN and ASTM [2]. This method has the ability to produce high-resolution parts with internal passages [3]. The AM technologies do not need any kind of tooling and it is possible to build undercuts. As a result, complex structures can be realized with these methods [4]. New lightweight constructions or individual parts can be realized. Large-volume parts or mass accumulation needed for the casting process can be transformed into framework structures. In this case, the mechanical properties (e.g., stiffness, strength) will be adapted by the framework structure. Ahmadi et al. investigated different structures with regard to their mechanical properties. Their investigations showed that the truncated cube and the truncated cuboctahedron are two of the strongest structures, while the diamond and rhombic dodecahedron structures are comparatively weak. The same behavior was observed for the stiffness. The stronger structures also have a higher stiffness [5]. Additionally, the microstructure strongly depends on the production method. Relating to Trosch et al., as-cast Inconel 718 parts have grain sizes at the scale of 1 mm. With forging, the grain size can be downsized to 10–100 µm. However, with PBF-LB, a grainsize of 1–5 µm can be realized. Additionally, the grain size can be tailored by adjusting the laser parameters. Increasing the scanning speed, for example, leads to a more refined grain size [6]. According to the Hall–Petch relationship, a finer microstructure results in an increased yield strength from 940 (as-cast) to 1185 MPa (PBF-LB) for Inconel 718 [7]. This mechanism translates to other metals and alloys as well and is especially relevant for the strengthening of magnesium alloys.

The PBF-LB process poses several challenges itself. Like the as-cast products, parts manufactured using PBF-LB have residual porosity, and the solution to this issue is dependent on the exact manufacturing process being employed. Kelly et al. optimized the density of Ti6Al4V parts by varying the laser parameters from 96.2% to 99.2%. Thus, the compressive yield strength and compressive yield strain of gyroid structures could be enhanced by 70% and 21%, respectively [8]. Another particularly important challenge with metal powders for AM is the oxide layer present on the powder particles. Due to the high specific surface area of the powder, the proportion of the metal oxide is higher than in other production methods that utilize bulk workpieces. Oxides can prevent the fusion of the particles, especially if they will not melt at the process temperature being used [9]. Significantly higher boiling points of metals in comparison with the melting points of their respective oxide lead to a better processability. The boiling point of metallic titanium (Ti) at 3287 °C is substantially higher than the melting point of the oxide layer (1842 °C). This is similar for iron with a boiling point of the metal of 2861 °C [10]. Mg is problematic in this regard as it has a low melting point of 650 °C and a boiling point of 1093 °C, combined with the oxide's very high melting point of 2825 °C [10]. This problem can be dealt with by either changing the material composition by alloying or adjusting the laser process itself. In previous investigations, the commercial alloy WE43 emerged as a promising alloy in producing Mg parts with PBF-LB [11]. WE43 consists nominally of magnesium alloyed with 4 wt.% yttrium (Y) and 3 wt.% rare-earth elements (RE). These elements are able to thermodynamically reduce the thermally very stable magnesium oxide (MgO) [10]. In addition, certain rare-earth oxides form an eutectic system with MgO, which has the potential to reduce the melting temperature of the oxide layer by 800 °C [12].

The first investigations to fabricate single tracks out of magnesium using PBF-LB were conducted by Ng et al. (2009) [13]. They proved the feasibility of this process, despite the difference in the evaporation temperature of the oxide compared to the melting temperature of pure magnesium. They also found that the process is highly dependent on the processing parameters, especially the laser power and laser scanning speed. There was an unstable melting process, with much spattering and process emission, diminishing the build quality. In order to improve the process, many investigations were carried out over the next few years to adjust the process parameters for the fabrication of ideal single tracks and completely dense samples [14,15]. Jauer et al. produced the first fully dense sample in 2016. They used a dedicated laser melting system, designed to remove process emissions out of the building chamber. Moreover, they used the magnesium alloy AZ91 to realize samples with a relative density up to 99.5% [16]. However, the problem with aluminum as an alloying component is its poor biocompatibility, which led to increased process development activities for the use of WE43, which has been used in conventionally manufactured resorbable implants previously [15,17].

The objective of the present study was the comparison of the properties of products generated with either PBF-LB or die casting. Hence, the densities, compositions, microstructures and mechanical strengths of the products were investigated.

2. Materials and Methods

2.1. Magnesium Alloys

In this study, the Mg alloy WE43 (4 wt.% Y and 3 wt.% rare-earth metals, balance Mg) was investigated. For the production of the as-cast samples, WE43 ingots were provided by Magnesium Elektron UK (Manchester). In the PBF-LB experiments, a gas-atomized WE43 powder provided by Carpenter Additives (Widnes, Cheshire, UK) was used. An SEM (scanning electron microscope) image showing a typical selection of powder particles is displayed in Figure 1. The particle diameters are below 63 µm with a median of 9.6 µm and a 0.9 quantile at 18.7 µm. The shapes and the sizes of the powder particles are very inhomogeneous, having, in part, adhering satellites and deformed particles. With the 90% quantile at 18.7 µm, the manufacturing of layer thicknesses down to 20 µm is possible. Moreover, the powder has excellent flowability, which is important even for powder application, and therefore a homogenous melting behavior.

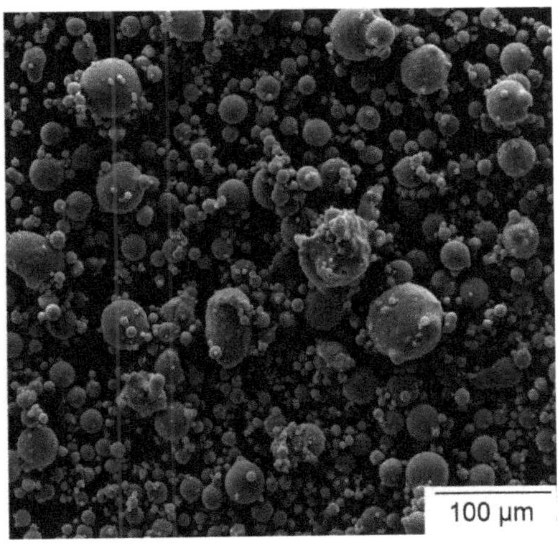

Figure 1. SEM image of the WE43 used for PBF-LB.

2.2. Casting of the WE43 Samples

The reference samples were cast using a gravity die casting process. For the casting process, a mild steel crucible and steel dies (both self-built) were used. The casting geometry inside the dies was a horizontal rod (∅ 20 mm × 220 mm) with a feeder along the top of the rod. Among other things, this geometry was also designed to obtain a homogeneous microstructure. The properties of the samples should consequently be homogeneous along the rod. To prevent interfacial reactions or excessive diffusion of iron into the melt, the crucible was coated with boron nitride (BN). For the die, two different coatings were used. The rod was coated with graphite for a faster cooling rate and the feeder was coated with BN to slow down the cooling rate. The WE43 ingots were then placed into the crucible. For the melting process, a resistance-heated furnace (Nabertherm GmbH, Lilienthal, Lower Saxony, Germany) was used. To shield the Mg melt from oxidation, a constant gas stream of 1000 mL/min of 0.3 vol.% reactive SF_6 in nitrogen over the top section of the crucible was maintained. The WE43 ingots were melted and the temperature of the melt was then raised to 750 °C. Simultaneously, the dies were preheated to 350 °C. After reaching the casting temperature, the crucible was taken out of the furnace and the melt was cast manually into the preheated dies. Finally, the feeder and sprue were cut off mechanically after demolding.

The samples for the characterization were prepared out of the as-cast rods (Knuth Wergzeugmaschinen GmbH, Wasbek, Schleswig-Holstein, Germany and EMCO GmbH, Hallein-Taxach, Salzburger Land, Austria). The microstructure of the whole cross-section was analyzed on cylindrical samples (∅ 20 mm × 10 mm) cut out of the rods. The cylindrical samples for the compression tests had a diameter of 5 mm and a length of 7.5 mm. The tensile test specimens had the B4 ×20 sample size according to the standard DIN 50125 and the flat samples for measuring the bending strength were cuboid with dimensions of 43 mm × 6 mm × 2 mm. All these samples were machined out of the rod.

2.3. *PBF-LB of WE43*

In PBF-LB, a laser melts a multitude of particles in a powder bed, forming layers with a set height. After the melting process, the next layer is applied by first lowering the build platform, applying the powder and melting the layer. This process is repeated until the desired geometry is built. In this process, several parameters have a substantial impact on the resulting geometry and the mechanical properties. A parameter study was executed to establish suitable process parameters for fabrication of the samples. The following parameters were varied: laser power, laser scanning speed, hatch distance, hatch pattern, build plate preheating and layer size.

The samples in PBF-LB were fabricated with a laser melting system of the type SLM125HL (SLM Solutions Group AG, Lübeck, Germany). The specimens were cubes with a size of 5 mm × 5 mm × 5 mm with a 1 mm support structure. The generated samples were analyzed with the statistical software JMP (Version 15, SAS Institute, Cary, NC, USA), in order to gather information in a wide processing interval with the smallest sample possible. A series of three investigations was carried out to determine the processing parameters for the highest relative density with the lowest respective porosity. With the parameters and intervals shown in Table 1, a statistical design of the experiment was developed with the software JMP. The parameter screening was based on a central composite design, which was divided into three parts. The first experiment with 18 sets of parameters was a screening to narrow the possible parameters down to a highly relevant selection. In the following, the best laser parameters were chosen by 36 additional experiments. The best parameters of those experiments were then used for a more detailed analysis in another six additional experiments.

Table 1. Parameters for the central composite design of the first parameter screening.

Step	Process Parameters for Screening Experiment					
	Laser Power in W	Scanning Speed in mm/s	Hatch Distance in μm	Layer Height in μm	Hatch Pattern	Build Plate Temperature in °C
−1	20	100	10	20	Lines	40
0	60	800	80	-	-	-
1	100	1500	150	75	Chess	200

The objective of this process development was to create an empirical process model of WE43 for the powder bed fusion. This model predicts the relative density from a combination of laser and manufacturing parameters, is calculated with the software JMP and consists of numerous embedded equations. These are based on a selection of parameters in individual equations to show the individual influence and are furthermore crossed to take the interactions between them into account.

2.4. *Mechanical Testing*

The bending strength was determined with three-point bending tests according to the standard DIN EN ISO 7438 using a Zwick Z250kN mechanical testing machine (ZwickRoell GmbH, Ulm, Germany) with a 20 kN load cell. The samples were placed on a two-point bearing surface with a distance of 17 mm. The testing started with the force set to 10 N, followed by the actual bending test with a crosshead speed of 1 mm/min.

The compression tests were performed according to the standard DIN 5016 using a Zwick Z 250 Retro Line mechanical testing machine (ZwickRoell GmbH, Ulm, Germany).

The preload used in the compressive test was set to 50 N, followed by a nominal strain rate of 10^{-3} s^{-1}.

Furthermore, the behavior of the as-cast samples under tensile load was characterized according to the standard DIN 50125 with the Zwick Z 250 Retro Line mechanical testing machine and a 20 kN load cell. The preload used in the tensile test was set to 25 N.

For each investigated parameter, a minimum number of eight samples were tested to statistically validate the results of the mechanical tests. For the parts generated with AM, the mechanical properties of two separate build jobs were characterized for an indication of the overall reproducibility of the process.

In addition, the Vickers hardness was analyzed according to the standard DIN EN ISO 6507-1 using a Zwick ZHU 250 (ZwickRoell GmbH, Ulm, Germany). The method HV10 was employed using a Vickers indenter and a test force of 98.1 N. On each sample, the average hardness was determined by calculating the mean values of at least seven indentations.

2.5. Microstructural Analysis

The as-cast magnesium implants were ground and polished down to a particle size of 1 µm of the polishing agent. Afterwards, the surface underwent a treatment with 10% picric acid in ethanol in order to slightly etch the grain boundaries. The images of the microstructure were taken with a VK-X1000 microscope (Keyence, Neu-Isenburg, Germany). The grain size analysis was carried out with different methods for the as-cast and additive manufactured samples. The microstructure of the parts generated with PBF-LB showed differently shaped grains. To get a meaningful result, the areas of the individual grains were determined. Therefore, the grain boundaries were marked separately to calculate the inner areas. The as-cast samples, however, had mostly globular grains and hence the grain sizes of the as-cast samples were analyzed according to the ASTM E 112 standard. Using the calculated grain size, the occupied area of the grains was calculated with the assumption that the grains were spherical. The scanning electron microscope (SEM) images of the microstructure and the energy-dispersive X-ray spectroscopy (EDX) analysis were obtained with a SUPRA 55 VP (Carl Zeiss AG, Oberkochen, Germany) using an acceleration voltage of 12 kV, a spot size of 2.5 nA, a working distance of 6.2 mm and magnifications of 50×, 250× and 400× (as-cast) and 50×, 500× and 6000× (PBF-LB).

2.6. X-ray Characterization

Phase determination was carried out by X-ray diffraction (XRD) using a C8 Discover (Bruker Corporation, Billerica, MA, USA) with CuKα radiation (λ = 1.540562 Å). The samples were scanned continuously from 15° to 88° in 2θ with a step size of 0.01°.

The porosity of the as-cast samples compared with the AM-generated parts with different process parameters was investigated with an Xradia 520 Versa X-ray microscope (Carl Zeiss AG, Oberkochen, Germany). The images were taken with a CCD camera (Andor Technology Ltd., Belfast, Northern Ireland, UK) and lenses at 4 times magnification. For the as-cast samples, a spatial resolution of 2.4 µm, an acceleration voltage of 70 kV, a power of 6 W and an exposure time of 3.2 s were used. The tomographic images of the parts generated with AM were implemented with spatial resolutions of 5 (samples for the compression test) and 6 µm (samples for bending tests), an acceleration voltage of 80 kV, a power of 7 W and exposure times of 5 (samples for the compression test) and 8 s (samples for bending tests).

3. Results

3.1. Process Development

The objective of the process development was to create a process model to be able to determine a set of parameters adjusted to the fabrication of high-quality WE43 parts. This process model links the laser and manufacturing parameters via empirical equations to the relative density as the target variable. Figure 2a–c show three specimens in a cross-section view, displaying the influence of the energy input by the laser on the relative density.

These cross-sections as a base for analysis are oriented perpendicular to the manufactured layers to obtain detailed information about the course of the building process. This allows faults of the melting behavior to be detected in individual layers and to, furthermore, be correlated with laser exposure strategies. The corresponding process parameters are given in Table 2. In addition, the set of parameters given by the model for a maximum density, which was also used to process the samples in the present study, was added in Figure 2d and as #d in Table 2. It is shown that the forecast of the process model is valid and the calculated parameter set for maximum density leads to 99.9% density.

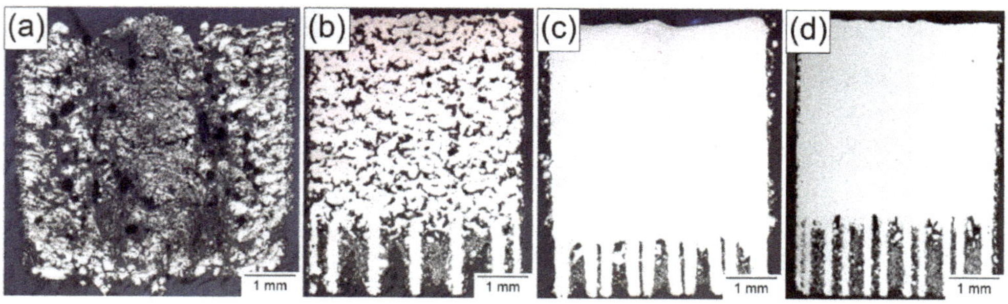

Figure 2. Cross-sections of representative specimens from the process screening (**a–c**) with the parameters shown in Table 2; best overall result of this process development based on the final empirical process model (**d**).

Table 2. Set of parameters used for processing of the specimens shown in Figure 2.

Option	Process Parameters							
	Laser Power in W	Scanning Speed in mm/s	Hatch Distance in μm	Layer Height in μm	Hatch Pattern	Build Plate Temperature in °C	Energy Input in J/mm³	Relative Density in %
a	20	100	10	20	Lines	200	1000	42.2
b	60	800	80	75	Lines	200	12.5	80.0
c	100	800	10	75	Chess	200	625	99.9
d	80	450	45	20	Chess	40	197.5	99.9

The specimen shown in Figure 2, option a, clearly had an energy input which was too high. If the energy input is too low, dense samples could not be realized either, as is seen in Figure 2, option b. A correct energy input leads to dense samples with a porosity of less than 0.1%. Figure 2, option c, corresponds to the best result of the parameter screening and had 99.9% relative density. This specimen had sufficient density, but showed poor dimensional accuracy. The layer height of 75 μm is desirable due to the good build-up rate but results in a significant stair chase effect on angled surfaces. Thus, a parameter with a high density for a smaller layer height had to be identified, and the statistics software was used to create an empirical process model. To simplify the model and the calculation, some processing parameters were fixed to eliminate variables. The parameters were set to 20 μm layer height, 40 °C build plate preheating and chess hatching with a field size of 4 mm edge length as a laser exposure with a 90° hatch rotation from layer to layer. With these parameters as a basis, the next experiment investigated the influence of the laser power, scanning speed and hatch distance in detail. The selection was made due to experience in preliminary work, where these three variables had the most significant impact. Processes could be optimized to very good results, disregarding the layer height, preheating temperature and laser exposure strategies due to their minor influence. The laser power was varied between 20 and 100 W, the scanning speed from 100 to 1500 mm/s and the hatch distance from 10 to 150 μm. The parameters used were based on the forecast of the process model.

In this parameter interval were six parameter combinations, which led to relative densities of 99.9%. The best result is #d in Table 2, which showed a very high density and a good dimensional fit. This parameter set was then used for the fabrication of the samples for the mechanical testing as well as the microstructure investigation.

3.2. Tomographic Analysis

The spatial distributions of the rare-earth precipitates and the pores were determined by volumetric XRM scans. Typical cross-sections from two different planes of cast and additively manufactured parts are shown in the 2D slice images in Figure 3. In the as-cast samples (Figure 3a,b), no pores were detected, which means the pores are smaller than the detection limit or no pores exist in the investigated area. In addition, the rare-earth-containing phases in the microstructure, which are the brighter spots in the tomographic images due to higher X-ray attenuation, are in the micrometer range with only a few agglomerated spots of those rare-earth precipitates. Clearly, it was demonstrated that the precipitates have no preferential orientation. In comparison, pores were detected in every sample of the parts generated with PBF-LB (Figure 3c–h). The cylinders for the compression tests had smaller spherical or elongated pores, which were arranged between the layers. The size was in the range of 20 to 100 μm, whereas elongated pores reached a length of up to 160 μm. Considering the precipitates, those were also larger than in the as-cast samples and were arranged in layers. Rare-earth precipitates emerged preferably at grain boundaries [18]. Thus, smaller and uniformly distributed precipitates may have resulted from smaller grains. Considering the cuboid-shaped PBF-LB samples for the bending test, the porosity varied strongly with the building process, although the same parameters were used. In the first process, the porosity was low and similar to the smaller samples for the compression test. By producing the same cuboid-shaped samples in a second process, the porosity increased strongly. However, only the amount of pores increased; their size remained the same. In addition, the layered structure of the process was visible based on the distribution of the precipitates. The porosity strongly affected the mechanical behavior as will be demonstrated later.

3.3. Microstrucure and Phase Analysis

The microstructure of the WE43 specimens strongly depended on the manufacturing process, as is shown in Figure 4. In the as-cast state (Figure 4a), the grains were comparatively large with mostly globular shapes and a mean grain size of 22 ± 0.5 μm. To be able to compare the grain size of the as-cast state with the PBF-LB state, the area was calculated on the assumption that the grains are spherical. In this case, the grains had an area of 380 μm^2. Unlike the porosity, the microstructure of all parts generated with PBF-LB was similar, which is why only representative images of the individual areas are shown in the following. The microstructure of the PBF-LB parts must be considered at two levels. At a mesoscopic level, the structure consists of parabolic areas with different sizes (Figure 4b). The parabolic structure was due to the melting of small separate spots and shows the direction of solidification. Each melt pool solidified in the same direction that the laser scanned along the surface. If pores occurred, they appeared mostly at the borders of the parabolic structures. This also indicates that the scanning parameters can influence the porosity strongly. In contrast to the as-cast condition, the grains of the PBF-LB parts consisted of much smaller and arbitrarily shaped grains. In the middle of the parabolic structures, the grains were more spherical (Figure 4c), while the grains on the borders of the parabolas could be both spherical or elongated (Figure 4d), whereby the latter grains were more pronounced and were visibly larger. Due to the elongation of the grains, a meaningful average grain size could not be determined using the same procedure as for the as-cast samples.

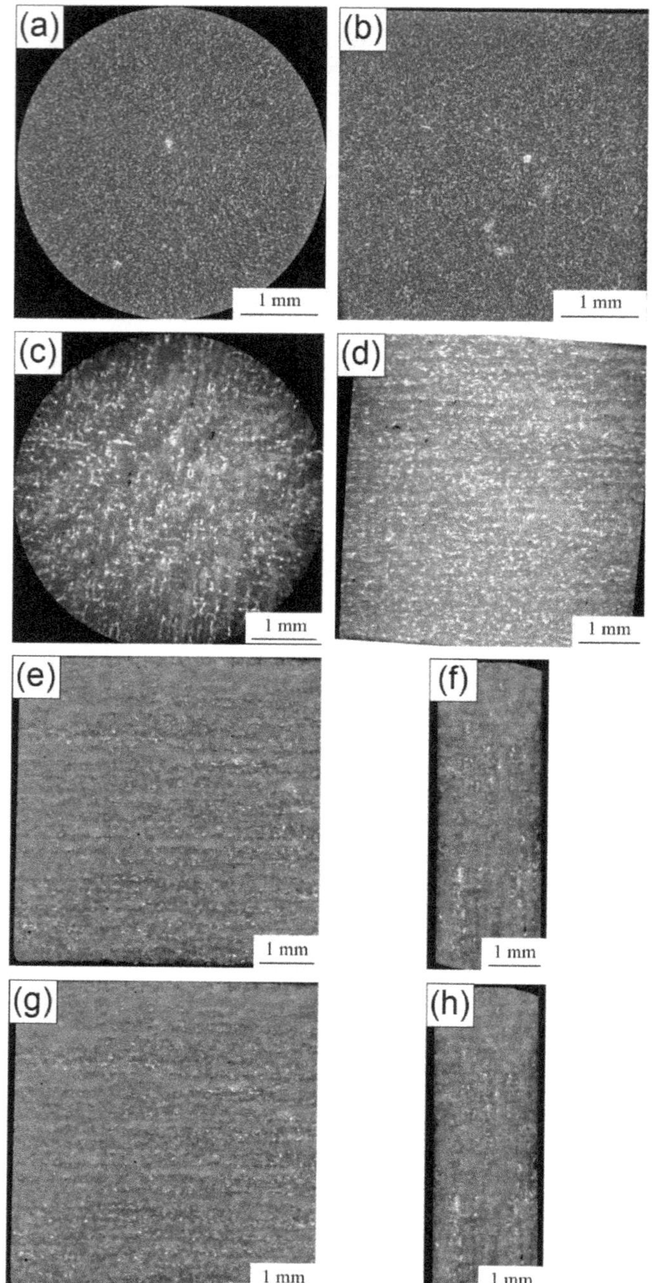

Figure 3. 2D images from the tomographic measurements of (**a**,**b**) as-cast WE43, (**c**,**d**) PBF-LB-generated WE43 samples for the compression test, (**e**,**f**) first PBF-LB-generated sample for the bending tests and (**g**,**h**) second PBF-LB-generated sample for the bending tests.

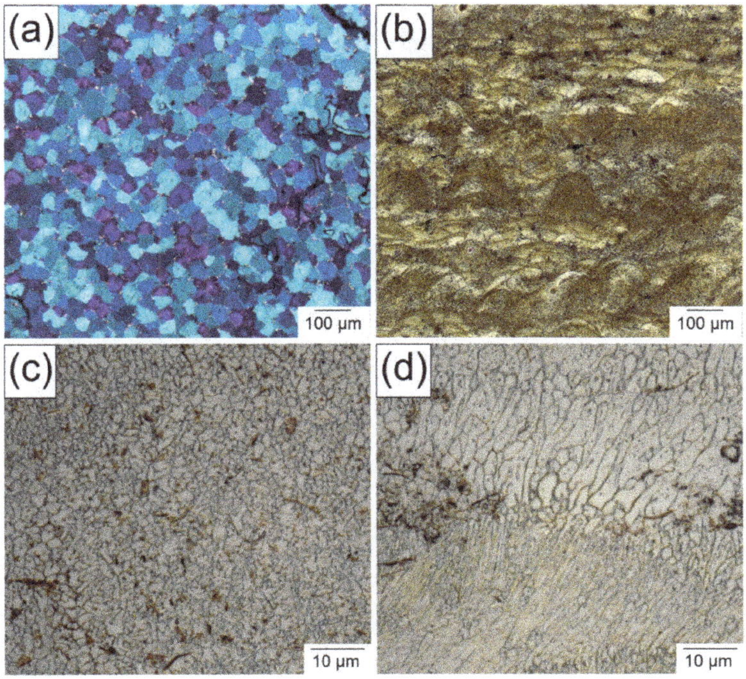

Figure 4. Light microscopy images of the microstructure of WE43: (**a**) as-cast, (**b**) PBF-LB-generated and (**c**,**d**) detailed images of different spots in (**b**).

Thus, the surface of the grains was analyzed and plotted in a histogram (Figure 5). Concluding from the numeric proportion, it is apparent that a majority of the grains were spherical ones in the micron or submicron range, as shown in Figure 4c. Towards the larger grains, the numeric histogram flattened out slowly due to the elongated grains displayed in Figure 4d. Although the amount of smaller grains is significantly higher, the area proportion of the elongated grains was similar to that of the spherical grains, as represented by the surface area histogram. Hence, the two grain geometries in Figure 4c occurred with an equal share of the surface area.

The different phases and their composition were characterized in SEM images along with additional EDX measurements. The backscattered electron images of the as-cast WE43 with the additional spots of the EDX measurements are shown in Figure 6 with a 50× (Figure 6a) and 250× (Figure 6b) magnification, together with the compositions of the different areas in Table 3. As the backscattered electrons show an element contrast, the distribution of the rare earths is qualitatively visualized. The overview image shows a needle-like structure of the Mg matrix, separating areas with a higher amount of rare-earth metals. In addition, precipitates, visible as small bright spots, were uniformly distributed. Both the precipitates and the separated parts in the matrix were homogeneously distributed. With the EDX measurement, only neodymium (Nd) could be detected out of the elements of the rare-earth mixture and had the expected magnitude of 2.4 wt.%. In contrast, the yttrium (Y) content was almost double the target value, with 7.9 wt.%. However, the tested volume was comparatively small, which is why an additional spark spectrometer measurement was employed. In this way, a larger volume of the sample was measured and the data were close to the nominal concentrations. Looking at the detailed image (Figure 6b), the center of the needle-like structure was in the middle of the grains. It was also found that most of the Nd was in the precipitates (EDX 4), whereas yttrium was concentrated in the separated part of the matrix. These results were confirmed for the overall structure through

EDX mappings (Figure 7) of the section in Figure 6b. The Nd content in all precipitates was found to be the same as was measured on a single precipitate (EDX 4). In contrast, Y did not show a significant increase in the precipitates compared to the Y found in the matrix. Due to the higher resolution of the SEM compared to the XRM, the pores, which are represented by the black dots in Figure 6a, could be revealed.

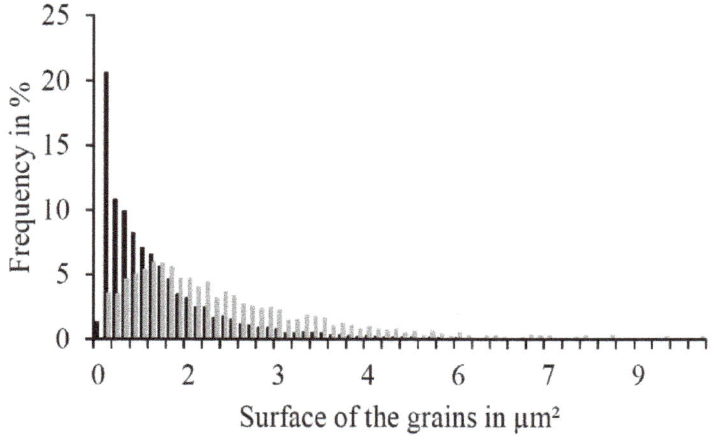

Figure 5. Histogram of the area-based grain size of PBF-LB-generated WE43 of the spherical grains (**gray**) and longitudinal grains (**black**).

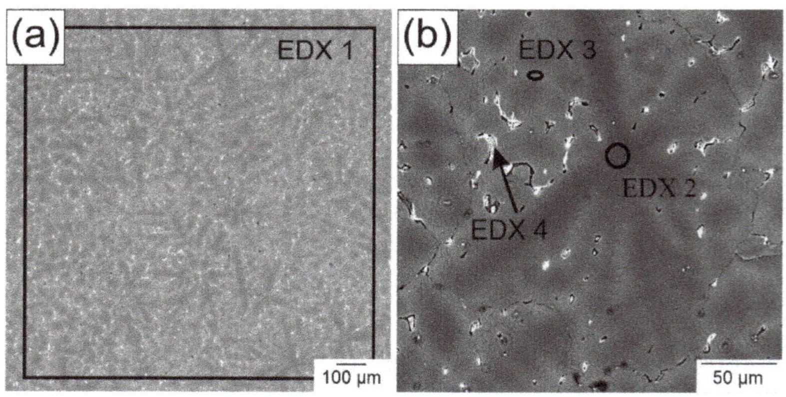

Figure 6. Backscattered electron images of as-cast WE 43 with (**a**) low and (**b**) high magnification.

Table 3. Results of the EDX measurements indicated in Figure 6.

Element	Composition in wt.%				
	EDX 1	**EDX 2**	**EDX 3**	**EDX 4**	**Spark Spectrometer**
Mg	89.7	97.8	94.9	84.0	94
Y	7.9	1.0	3.0	4.7	3.9
Nd	2.4	1.2	2.1	11.3	2.1

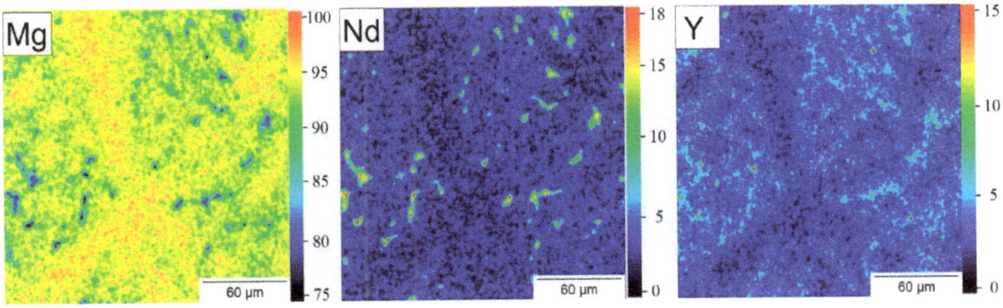

Figure 7. Element mapping of Mg, Nd and Y of the specimen in section in 6b.

Unlike the grain structures of the parts generated with PBF-LB, the phases of these parts were arranged differently. In the samples for the bending tests (Figure 8a,b), most of the precipitates were inhomogeneously distributed. Only a few layers consisted of evenly distributed precipitates. The higher-magnification image (Figure 8b) shows that the precipitates were agglomerated in some areas, but they still formed mostly fine structures in those agglomerations. Only in some cases did clusters of precipitates occur. In contrast, large parts of the compression test samples (Figure 8c,d) contained small and homogeneously distributed precipitates. However, there were still areas with agglomerated precipitates similar to the cuboid samples, as shown in Figure 8e,f. Although, locally, a homogeneous structure could be realized, the process was still not stable enough to produce a large volume with a homogeneous structure and a density above 99%. As with the as-cast samples, the composition of the different phases was also analyzed in the PBF-LB-generated parts. Besides the spots in Figure 9, a large-area EDX measurement (EDX 5) of Figure 8a was conducted. The compositions of the different phases, independent of the samples, were similar to each other, which is why only a representative EDX analysis of one PBF-LB-generated part is shown in Table 4. The Nd and Y content is rather high in the large-area EDX measurement. To determine the impact of the process on the composition, an additional EDX measurement of the initial powder was conducted on several particles. The RE content in the initial powder was actually found to be lower than in the final produced parts. As was mentioned before, the problem with Mg in the PBF-LB process was the low boiling point of metallic Mg. During the heating of each spot, Mg could have been vaporized to a greater extent than during the casting process.

In the matrix, there was also a variation in the RE content similar to the as-cast sample. Due to the finer microstructure, the separate parts were thinner. Hence, an accurate composition was hard to obtain because the excitation volume could easily include RE-rich areas. Consequently, the measured RE content of the matrix is much higher than the solubility limit. In contrast to the as-cast sample, the Y content in the precipitates was remarkably raised instead of the Nd content. According to the phase diagrams, Nd-rich precipitates should occur. However, the cooling rate in the PBF-LB process was much faster. The resulting intermetallic phase of Nd and Mg has a lower diffusion rate and could not be separated into a second phase. Thus, the formation of Nd-rich precipitates is kinetically inhibited.

The XRD analysis shows the occurring phases in the PBF-LB and as-cast samples, c.f. Figure 10. As expected, the Mg peaks were the most distinctive ones. Besides Mg, the phase Mg_3Nd occurred in samples of both production methods, although those precipitates were not found in the SEM/EDX analysis of the PBF-LB-generated parts. Therefore, the formation of Mg_3Nd was not kinetically inhibited, but rather had a diffusion-based formation of the precipitates. Furthermore, also striking is that only the oxide of yttrium (Y_2O_3) could be detected, especially in the PBF-LB parts, which is most prominent in the peak at 29°. Those distinctive peaks could not be detected in the as-cast samples. Consequently, the formation of oxides was process-based and did not emerge after the preparation of the samples. The

Y-rich precipitates in the PBF-LB samples could consequently be Y_2O_3. Unfortunately, the oxygen content could not be quantified with the EDX measurement before, which is why the precipitates of Y_2O_3 could not be localized by the SEM-EDX analysis. However, only Y_2O_3 was found as a Y-rich phase in the PBF-LB-generated parts. Consequently, those precipitations in Figure 9 should be Y_2O_3. This originates from the comparatively high amount of oxides in the powder-based process, as it was mentioned before.

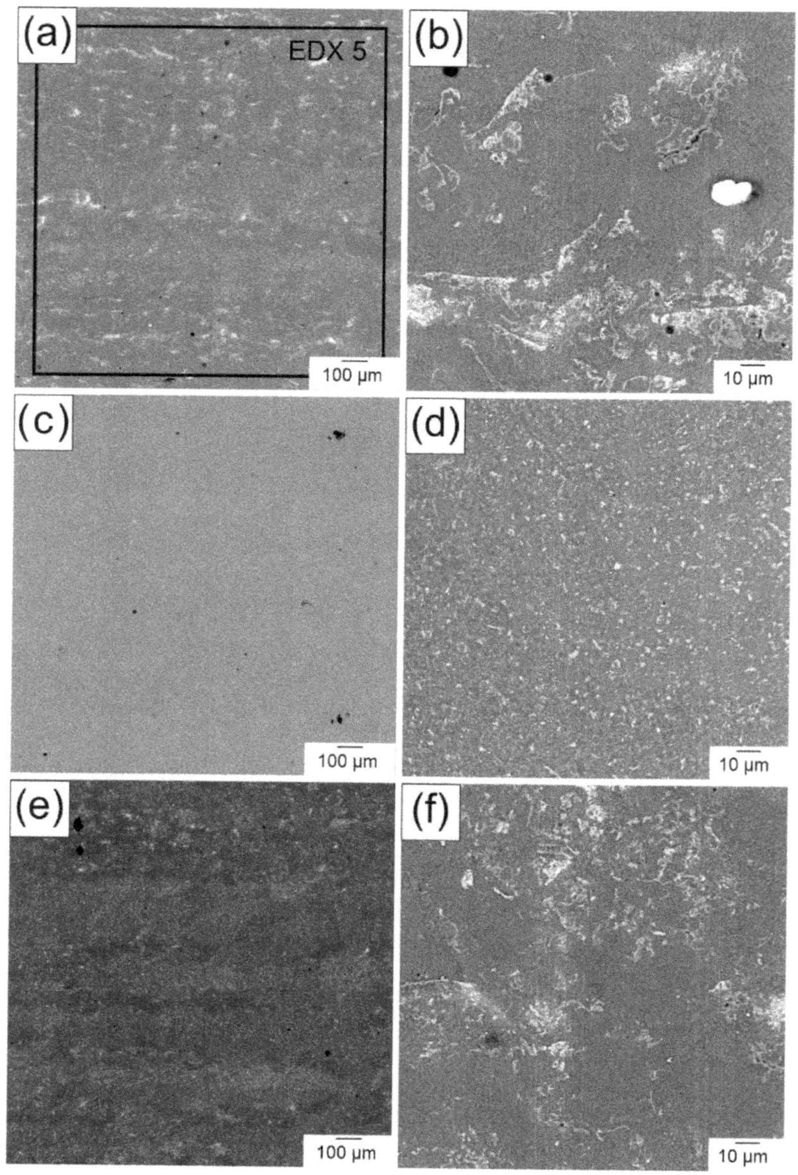

Figure 8. Backscattered electron images of PBF-LB-generated WE43: (**a**,**b**) sample for the porous bending test and (**c**–**f**) sample for the compression test.

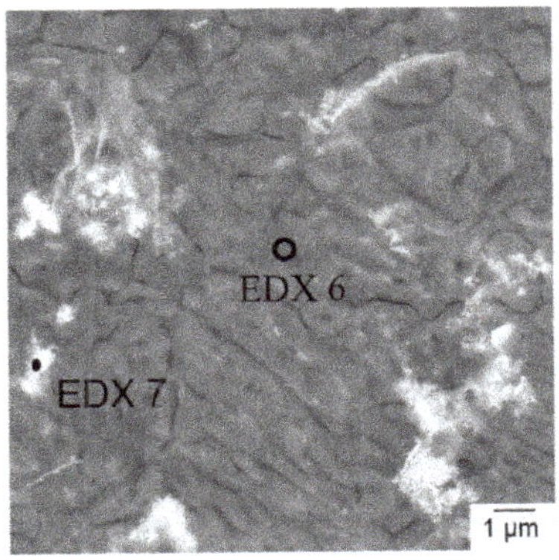

Figure 9. Backscattered electron images of PBF-LB-generated WE43 with the areas of the EDX measurements.

Table 4. Composition of the EDX measurement of the spots in Figures 8 and 9.

Element	Composition in wt.%			
	EDX 5	**EDX 6**	**EDX 7**	**WE43 Powder**
Mg	89.4	91.5	73.0	91.6
Y	6.2	4.4	22.1	4.6
Nd	4.4	4.1	4.9	3.8

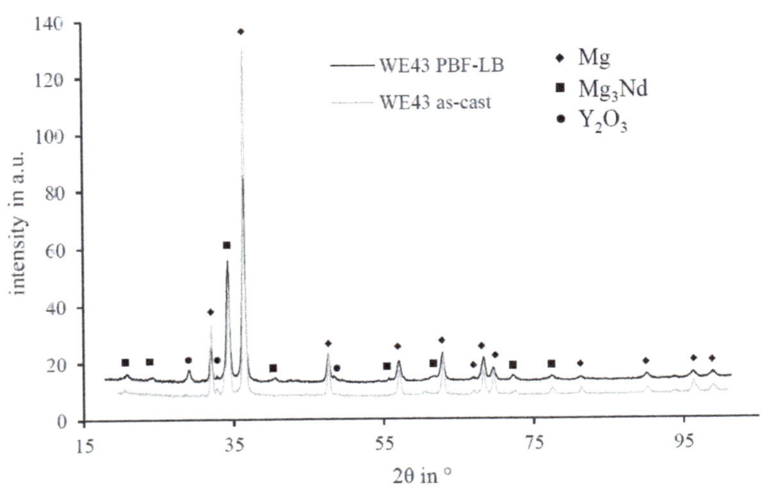

Figure 10. XRD pattern of as-cast and PBF-LB-generated WE43 with the corresponding phases.

3.4. Mechanical Properties

The mechanical properties are a key factor for lightweight constructions, especially for AM-generated parts, as they can be designed without considering the manufacturability by casting or forming. On this account, the as-cast and PBF-LB-generated parts were tested under compressive and bending stresses. However, up until now, the behavior under tensile stress was only tested with the as-cast samples. As it was mentioned before, the building direction of the samples for the mechanical testing generated by PBF-LB was along the weakest direction. The stress–strain diagrams of the behavior under compressive and bending loads are shown in Figures 11 and 12, respectively. The corresponding characteristic values of the compressive, bending and tensile tests and the hardness are listed in Table 5. The elastic behavior of WE43 does not depend on the production method, i.e., parts produced by PBF-LB and gravity die casting had similar elastic moduli. However, the elastic modulus is a physical quantity and depends on the bonding strength of the atoms, whereas the strength strongly depends on the microstructure, which was influenced by the production method. The as-cast samples had a comparatively low yield strength, which leads to an early onset of plastic deformation. Plastic deformation of PBF-LB-generated WE43 parts started at double the applied stress when compared to the as-cast parts. The compressive strength of PBF-LB-produced parts was also higher, but the difference was less pronounced. However, the ductility of the as-cast parts was higher as compared to PBF-LB, as shown in Figure 11. In fact, the elongation to fracture was doubled when using the as-cast parts. The samples showed little variation in mechanical behavior, so only one representative stress–strain diagram is represented for each case in Figure 11.

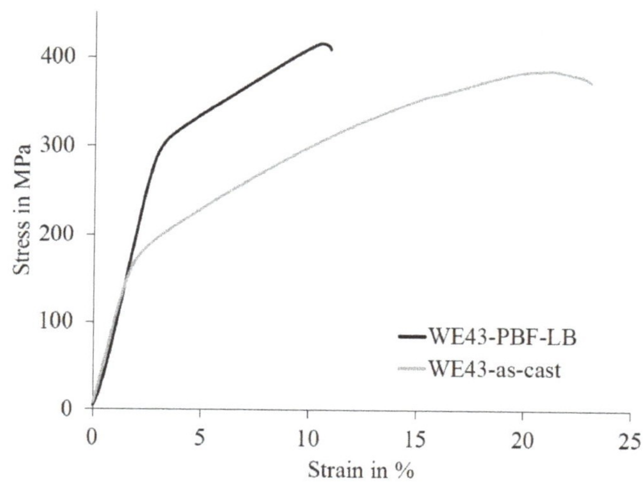

Figure 11. Stress–strain diagram of as-cast and PBF-LB-generated WE43 under compressive load.

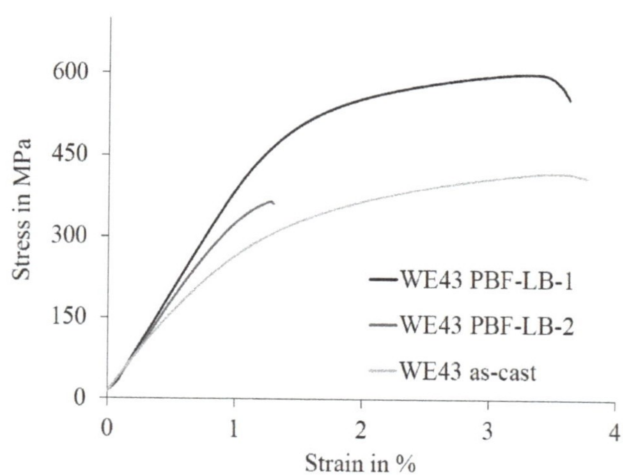

Figure 12. Stress–strain diagram of as-cast and PBF-LB-generated WE43 under bending load.

Table 5. Mechanical properties of as-cast and PBF-LB-generated WE43 parts.

Load Type	Characteristic Value	WE43 as-Cast	WE43 PBF-LB	
Compressive load	Yield strength in MPa	146 ± 7	297 ± 8	
	Compressive strength in MPa	383 ± 37	424 ± 41	
	Elongation in %	20 ± 2	11 ± 2	
Bending load	Yield strength in MPa	271 ± 35	499 ± 10	359 ± 18
	Bending strength in MPa	430 ± 30	601 ± 31	375 ± 24
Tensile load	Yield strength in MPa	142 ± 2	-	-
	Tensile strength in MPa	184 ± 22	-	-
	Elongation in %	2.3 ± 2	-	-
Hardness	Vickers hardness in HV10	69.0	94.1	

The elastic deformation under bending load was similar in both build jobs of the PBF-LB-generated parts. However, the plastic deformation behavior differed significantly. While the first samples generated had a high yield strength and, additionally, a high bending strength, the samples generated in the second manufacturing process clearly had a low proportion of plastic deformation, which also begins at a much lower stress. The preparation of the samples for the bending test was not yet reproducible. The bending behavior ranged from brittle to ductile. In particular, the ductility of the PBF-LB parts with a low porosity (WE43 PBF-LB-1) was more similar to the as-cast samples under bending load than under compressive load. However, the bending strength of the as-cast WE43 was considerably lower.

Magnesium typically shows a tension pressure anomaly [19], which was partly observed in the present study. The yield strength was similar in the as-cast WE43. Only the compressive strength strongly increased compared to the tensile yield strength by 108%. The biggest difference, however, was the elongation to fracture. It is expected that the strength of WE43 generated with PBF-LB under tensile stress should increase similarly to the compressive strength, but this is still to be investigated.

Strengthening of a material by a finer-grained microstructure typically results in an increase in hardness. The same behavior could be observed in the present study. As shown in Table 5, the WE43 parts generated by the PBF-LB process had an increase in hardness by 36% as compared to the as-cast condition.

4. Discussion

In the present study, WE43 parts of high quality could be successfully generated by employing PBF-LB. By further adjustments to the process, a density of over 99.9% could be achieved. The laser power, the scanning speed and the hatch distance had a big influence on the quality of the produced parts. So far, densities of at least 99% have been achieved when using WE43 for PBF-LB by generating small cubes with an edge length of 3 mm. If the dimensions of the samples were increased (edge length of 5 mm) or a more complicated geometry had been used, the density was observed to decrease to 95% according to Gieseke [11]. The process development resulted in a parameter set, which increased the density up to 99.9% for parts in the size range between 5 and 10 mm. However, when the dimensions were increased significantly to 50 mm, the porosity could not be set reproducibly. The reason is suspected to be a change in heat conductivity and can be mitigated by further investigation of processing parameters in the desired volume. Therefore, it is recommended to tailor the specimen in testing to a similar geometry for the desired applications. Since in a process development of a novel material, the feasibility has to be tested first, this work is the first step in this direction. As a consequence, the abstractability must be improved by a better understanding of the melting and solidification behavior. Using the same set of parameters in different building jobs leads to different porosities. In large volume parts, the large amount of pores should be homogeneously distributed because the set of parameters was equal throughout the whole sample [8]. The tomographic images of the PBF-LB parts show a deviating behavior. The pores are not homogeneously distributed in the whole volume, which suggests that the process is not stable throughout. Possible solutions are widening the process window by developing a new process variant or by tailoring the Mg alloy towards the process's requirements. Even in a stable process with a minimized porosity, the number of pores is still higher compared to the as-cast samples. However, the PBF-LB samples of different sizes showed the same porosity when the process was stable. This suggests that the porosity is independent of the size. In contrast, the porosity of as-cast samples strongly depends on the thickness of each part. Thicker parts show higher shrinkages, followed by an increase in the pore size and porosity [20].

The phase analysis had revealed that in both cases, the material consisted mostly of Mg with Mg_3Nd precipitates. This is contrary to the expectations based on the phase diagram because the thermodynamically stable phase for low amounts of Nd in Mg is MgNd [21]. Dealing with the oxide layer is a key issue for Mg alloys in the PBF-LB process. Keeping this in mind, the only oxide phase that emerged with a detectable level was Y_2O_3. In the case of a mixture of MgO and Y_2O_3, the melting point of the oxide layer should be between 2046 (eutectic temperature) and 2401 °C (melting point of Y_2O_3) [12]. Comparing the two XRD patterns, the issue of the oxide content was clearly demonstrated. The higher yttrium fraction in the precipitates of the PBF-LB parts was most likely caused by the oxide because no other yttrium-containing phases were detected. Comparing all the oxides of the alloying elements, Y_2O_3 is thermodynamically the most stable one. In fact, yttrium is able to reduce the oxides of the other alloying elements [10].

The grain structure of the PBF-LB parts is independent of the size or the quality of the build job. The parabolic structure, which has its origin in the beam traces, is strongly pronounced. This typical process-related structure could also be found in the work of Wei et al., who produced AZ91 parts by AM [22]. Whether and to what extent the beam traces and the layered structure appear strongly depend on the material and on the process parameters. Thus, in CoCrMo alloys, the parabolic structure is very prominent. In contrast, Ti alloys, for example, can prevent the formation of this layered structure due to grain growth across the layers [23], although it is possible to reach a trace structure with Ti6Al4V by adjusting the PBF-LB process with the disadvantage of a high porosity [24]. Comparing the grain size of the AM-generated parts with the as-cast parts, the difference in grain size is of two orders of magnitude, which is typical when comparing both processes. The same behavior was investigated by Trosch et al. by comparing the grain size of AM parts, formed parts and as-cast parts. In their study, the grain size was three magnitudes higher than the AM parts. Even the forged parts had 10 times larger grains [7]. Zumdick et al. produced WE43 parts through AM that had spherical-like grains with a diameter of 1.1 μm. They also described the parabolic structure, but not with the elongated grains found in the present study [25]. In as-cast samples, such elongated grains are typical with an anisotropic solidification from the mold wall to the center [26]. Similarly, the solidification in the PBF-LB parts is vertical to the parabolic line, as if it were one separated melt pool that solidifies inwards. Those separated melt pools also have spherical grains in the center emerging in a homogeneous nucleation, shown in the work of Dahle et al. [26]. In the present study, the volume share of both parts was roughly equal. Using the influence of the process parameters, the properties can be adjusted to the component's requirements.

This principle was used by Eifler in his work to adjust the mechanical properties and the corrosion behavior of the Mg alloy ZNdK100 by adjusting the microstructure with different extrusion parameters. The resulting microstructure varied from a completely recrystallized structure with only small grains to a partly recrystallized structure with a bimodal grain size distribution of deformed grains of the as-cast state and recrystallized grains. The strength, for example, could be raised to over 350 MPa [27]. The mechanical properties are particularly important for lightweight constructions. Seitz et al. investigated different extruded Mg alloys with the alloy LAE442, showing a maximum in the bending strength of 570 MPa and a compressive yield point of 150 MPa [28]. The as-cast WE43 samples had a similar compressive yield strength but a significantly lower bending strength. The strength of WE43 can also be increased by extrusion, as was performed by Dieringa et al. They reached a compressive yield strength of 261.5 MPa and a compressive strength of 420.4 MPa. However, the PBF-LB parts had even higher strengths than those from extruded LAE442 and WE43. This could be attributed to the even smaller grain size of the PBF-LB process compared to casting and extrusion, which increases the strength due to the Hall–Petch effect [7,29]. Using the process to vary the microstructure of PBF-LB parts as Thijs et al. did for Ti6Al4V, the properties of additive manufactured Mg could also be adapted to the application [24].

The problems with Mg stem from its difficult processability, which results from the thermophysical properties of Mg and MgO; therefore, the useful process window is strongly limited. By developing a new process and a new alloy tailored to the special properties of Mg, the process window could be widened, which could stabilize the process and increase the parts' properties.

5. Conclusions

The present study demonstrated that dense magnesium parts with relatively high volumes can be produced by PBF-LB with the alloy WE43. The main results can be summarized as follows:

1. With a process development targeted for maximum relative density, the process can be stabilized to generate large parts while ensuring a high density.
2. The microstructure consists of a bimodal grain size distribution with smaller spherical grains and larger elongated grains. Using the laser parameters, the microstructure could be adapted to directly adjust the properties of the generated part. Compared to the as-cast state, the grain size is one to two orders of magnitude lower, which explains the high strength even for WE43.
3. Still the porosity of the PBF-LB parts is higher than in the as-cast parts, which reduces the strength. Thus, there is still a great potential in the PBF-LB process. In further investigations, additional adjustments, such as those to the laser parameter or the atmosphere, have to be made to stabilize the process for Mg. Furthermore, an alloy adapted to the process could enhance the process capabilities.
4. The PBF-LB parts mainly consist of Mg, Mg_3Nd and Y_2O_3. The objective of reducing MgO to Mg by the rare-earth elements could be achieved. Due to the fast cooling rate, no Mg_3Nd precipitates could be formed. With additional heat treatments, precipitates of these intermetallic phases could be realized and additionally change the components' properties.

Author Contributions: Project administration, C.H., J.H., L.O., C.K. and H.J.M.; Supervision, C.H., J.H., L.O., C.K. and H.J.M.; Editing, C.H., J.H., L.O., C.K. and H.J.M.; Conceptualization, L.O., H.J.M.; Methodology, C.H., J.H., C.K.; Writing—original draft, S.J., A.A. and N.G.; Writing—review & editing, S.J., A.A. and N.G. All authors have read and agreed to the published version of the manuscript.

Funding: This research was supported by the German Research Foundation (DFG) within the project "Tailor made magnesium alloys for selective laser melting: Material development and process modelling" (grant No. 40948513) within the SPP2122. The Xradia 520 Versa (grant No. 316923185) used in this study was also provided by the DFG. The publication of this article was funded by the Open Access Fund of the Leibniz University Hannover.

Data Availability Statement: The data presented in this study are available on request from the corresponding author. The data are not publicly available due to privacy.

Conflicts of Interest: The authors declare no conflict of interest.

References

1. Roland Berger. Wohlers Associates, VDW, Additive Manufacturing 2013. 2013. Available online: www.rolandberger.com (accessed on 22 October 2019).
2. DIN Deutsches Institut für Normung e.V. *ASTM International, Additive Fertigung: Grundlagen—Terminologie 01.040.25; 25.030*; DIN-Norm: Berlin, Germany, 2018.
3. Frazier, W.E. Metal Additive Manufacturing: A Review. *J. Mater. Eng. Perform.* **2014**, *23*, 1917–1928. [CrossRef]
4. Salmi, A.; Calignano, F.; Galati, M.; Atzeni, E. An integrated design methodology for components produced by laser powder bed fusion (L-PBF) process. *Virtual Phys. Prototyp.* **2018**, *13*, 191–202. [CrossRef]
5. Ahmadi, S.M.; Yavari, S.A.; Wauthle, R.; Pouran, B.; Schrooten, J.; Weinans, H.; Zadpoor, A.A. Additively Manufactured Open-Cell Porous Biomaterials Made from Six Different Space-Filling Unit Cells: The Mechanical and Morphological Properties. *Materials* **2015**, *8*, 1871–1896. [CrossRef] [PubMed]
6. Zhang, C.; Luo, A.A.; Peng, L.; Stone, D.S.; Chang, Y.A. Thermodynamic modeling and experimental investigation of the magnesium–neodymium–zinc alloys. *Intermetallics* **2011**, *19*, 1720–1726. [CrossRef]
7. Trosch, T.; Strößner, J.; Völkl, R.; Glatzel, U. Microstructure and mechanical properties of selective laser melted Inconel 718 compared to forging and casting. *Mater. Lett.* **2016**, *164*, 428–431. [CrossRef]
8. Kelly, C.N.; Francovich, J.; Julmi, S.; Safranski, D.; Guldberg, R.E.; Maier, H.J.; Gall, K. Fatigue behavior of As-built selective laser melted titanium scaffolds with sheet-based gyroid microarchitecture for bone tissue engineering. *Acta Biomater.* **2019**, *94*, 610–626. [CrossRef]
9. Richard, H.A.; Schramm, B.; Zipsner, T. *Additive Fertigung von Bauteilen und Strukturen: Neue Erkenntnisse und, Morgan Kaufmann*; Springer: Berlin/Heidelberg, Germany, 2019.
10. Haynes, W.M. *CRC Handbook of Chemistry and Physics*, 95th ed.; CRC Press: Boca Raton, FL, USA, 2015.
11. Gieseke, M. *Entwicklung des Selektiven Laserstrahlschmelzens von Magnesium und Magnesiumlegierungen zur Herstellung von Individuellen und Bioresorbierbaren Implantaten*, 1st ed.; TEWISS: Garbsen, Germany, 2015.
12. Jiang, D.; Mukherjee, A.K. Spark Plasma Sintering of an Infrared-Transparent Y 2 O 3 -MgO Nanocomposite. *J. Am. Ceram. Soc.* **2010**, *93*, 769–773. [CrossRef]
13. Ng, C.C.; Savalani, M.; Man, H.C. Fabrication of magnesium using selective laser melting technique. *Rapid Prototyp. J.* **2011**, *17*, 479–490. [CrossRef]
14. Gieseke, M.; Kiesow, T.; Nölke, C.; Kaierle, S.; Palmer, T. *Rapid.Tech—Trade Fair and User's Conference for Rapid Technology*; Messe Erfurt: Erfurt, Germany, 2015.
15. Wessarges, Y.; Gieseke, M.; Kiesow, T.; Kaierle, S.; Overmeyer, L. (Eds.) *Selective Laser Melting of Magnesium Alloys for Biomedical Applications*; DAAAM: Nuremberg, Germany, 2016.
16. Jauer, L.; Meiners, W.; Vervoort, S.; Gayer, C.; Zumdick, N.A.; Zander, D. (Eds.) *Selective Laser Melting of Magnesium Alloys*; GE Global Research: Nishkana, NY, USA, 2016.
17. Tandon, R.; Madan, D. Emerging Applications Using Magnesium Alloy Powders: A Feasibility Study. In *Magnesium Technology 2014*; Alderman, M., Manuel, M.V., Hort, N., Neelameggham, N.R., Eds.; Springer International Publishing: Cham, Switzerland, 2016; pp. 21–25.
18. Wei, L.Y.; Dunlop, G.L.; Westengen, H. Age hardening and precipitation in a cast magnesium-rare-earth alloy. *J. Mater. Sci.* **1996**, *31*, 387–397. [CrossRef]
19. Xiang, C.; Gupta, N.; Coelho, P.; Cho, K. Effect of microstructure on tensile and compressive behavior of WE43 alloy in as cast and heat treated conditions. *Mater. Sci. Eng. A* **2018**, *710*, 74–85. [CrossRef]
20. Reikher, A.; Barkhudarov, M.R. *Casting: An Analytical Approach*; Springer: London, UK; Berlin/Heidelberg, Germany, 2007.
21. Villars, P. Inorganic Solid Phases. 2016. Available online: https://materials.springer.com (accessed on 19 March 2019).
22. Wei, K.; Gao, M.; Wang, Z.; Zeng, X. Effect of energy input on formability, microstructure and mechanical properties of selective laser melted AZ91D magnesium alloy. *Mater. Sci. Eng. A* **2014**, *611*, 212–222. [CrossRef]
23. Gibson, I.; Rosen, D.; Stucker, B. *Additive Manufacturing Technologies*; Springer: New York, NY, USA, 2015.
24. Thijs, L.; Verhaeghe, F.; Craeghs, T.; van Humbeeck, J.; Kruth, J.-P. A study of the microstructural evolution during selective laser melting of Ti–6Al–4V. *Acta Mater.* **2010**, *58*, 3303–3312. [CrossRef]
25. Zumdick, N.A.; Jauer, L.; Kersting, L.C.; Kutz, T.N.; Schleifenbaum, J.H.; Zander, D. Additive manufactured WE43 magnesium: A comparative study of the microstructure and mechanical properties with those of powder extruded and as-cast WE43. *Mater. Charact.* **2019**, *147*, 384–397. [CrossRef]
26. Dahle, A.K.; Lee, Y.C.; Nave, M.D.; Schaffer, P.L.; StJohn, D.H. Development of the as-cast microstructure in magnesium–aluminium alloys. *J. Light Met.* **2001**, *1*, 61–72. [CrossRef]
27. Eifler, R. Einfluss der Rekristallisation auf die Mikrostrukturentwicklung der Magnesiumlegierung ZNdK100. Ph.D. Thesis, Leibniz Universität Hannover, Hannover, Germany, 2018.
28. Seitz, J.-M.; Collier, K.; Wulf, E.; Bormann, D.; Angrisani, N.; Meyer-Lindenberg, A.; Bach, F.-W. The Effect of Different Sterilization Methods on the Mechanical Strength of Magnesium Based Implant Materials. *Adv. Eng. Mater.* **2011**, *13*, 1146–1151. [CrossRef]
29. Hansen, N. Hall–Petch relation and boundary strengthening. *Scr. Mater.* **2004**, *51*, 801–806. [CrossRef]

materials

MDPI

Article

Physical and Geometrical Properties of Additively Manufactured Pure Copper Samples Using a Green Laser Source

Samira Gruber [1,*], Lukas Stepien [1], Elena López [1], Frank Brueckner [1,2] and Christoph Leyens [1,3]

1 Fraunhofer Institute for Material and Beam Technology, IWS, Winterbergstraße 28, 01277 Dresden, Germany; Lukas.Stepien@iws.fraunhofer.de (L.S.); elena.lopez@iws.fraunhofer.de (E.L.); Frank.Brueckner@iws.fraunhofer.de (F.B.); christoph.leyens@iws.fraunhofer.de (C.L.)
2 Department of Engineering Sciences and Mathematics, Luleå University of Technology, 97187 Luleå, Sweden
3 Institute of Materials Science, Technische Universität Dresden, Helmholtzstr. 7, 01069 Dresden, Germany
* Correspondence: samira.gruber@iws.fraunhofer.de

Abstract: So far, copper has been difficult to process via laser powder bed fusion due to low absorption with the frequently used laser systems in the infrared wavelength range. However, green laser systems have emerged recently and offer new opportunities in processing highly reflective materials like pure copper through higher absorptivity. In this study, pure copper powders from two suppliers were tested using the same machine parameter sets to investigate the influence of the powder properties on the material properties such as density, microstructure, and electrical conductivity. Samples of different wall thicknesses were investigated with the eddy-current method to analyze the influence of the sample thickness and surface quality on the measured electrical conductivity. The mechanical properties in three building directions were investigated and the geometrical accuracy of selected geometrical features was analyzed using a benchmark geometry. It could be shown that the generated parts have a relative density of above 99.95% and an electrical conductivity as high as 100% International Annealed Copper Standard (IACS) for both powders could be achieved. Furthermore, the negative influence of a rough surface on the measured eddy-current method was confirmed.

Keywords: additive manufacturing; laser powder bed fusion; pure copper; short wavelength laser system; green laser; eddy-current method; electrical conductivity

Citation: Gruber, S.; Stepien, L.; López, E.; Brueckner, F.; Leyens, C. Physical and Geometrical Properties of Additively Manufactured Pure Copper Samples Using a Green Laser Source. *Materials* **2021**, *14*, 3642. https://doi.org/10.3390/ma14133642

Academic Editors: Bilal Gökce, Eric Jägle and Manfred Schmid

Received: 2 June 2021
Accepted: 22 June 2021
Published: 29 June 2021

1. Introduction

Copper has the second highest thermal and electrical conductivity of all non-superconducting materials, making it the material of choice for many functional applications such as heat exchangers or induction coils. Additive manufacturing processes offer a high degree of geometrical freedom for the fabrication of complex parts and are therefore a promising technology for pure copper applications. So far, pure copper has been processed via electron beam melting (EBM) [1–4], binder jetting (BJ) [5], laser powder bed fusion (LPBF) [6–14] and laser metal deposition (LMD) [15–17]. The highest relative densities of 99.95% and electrical conductivities of 96.24% IACS up until 2020 were achieved with the EBM process [2–4]. Bai and Williams [5] reached 97.3% relative density after sintering with the BJ process. Due to the high thermal conductivity of pure copper and low absorptivity in the range of 25% [11,18] of the copper powder at laser wavelengths of 1000–1100 nm, which are mostly used for LPBF [19], stable processing is not possible for infrared laser powers below 500 W and, therefore, fully dense parts could not be produced [13,20]. So far, only relative densities of 83–88% have been reported for pure copper parts when using a common 200 W infrared laser source [14,21]. One approach to increase the relative density of additively manufactured pure copper parts is by increasing the laser power. Colopi et al. and Ikeshoji et al. [12,22] have reached relative densities as high as 99.1% to 99.6% when using laser powers of up to 1 kW at a wavelength of 1 µm. However, melt-pool instabilities still occur mainly due to the different absorptivity of the powder bed and the molten copper at those wavelengths, leaving parts with low surface quality. By far the most promising approach is using a laser source with shorter wavelengths, primarily "green lasers" with wavelengths of about 515 nm. Such a setup can produce additively manufactured pure copper parts with relative densities of up to 99.8% and an electrical conductivity reaching up to 100% IACS [23], thus enabling the highest performances. A green laser source was already used for LMD [16] and even blue diode lasers showed promising results in dense single tracks [17].

The aim of this study was to verify the robustness of the process by using two powders with different particle size distributions and to take a closer look at the evolving density, microstructure, and effect of geometry features such as wall thickness and surface condition on electrical conductivity measurements. The geometrical accuracy was analyzed via 3D scanning and the mechanical properties were also determined for three building directions.

2. Experimental Setup

2.1. Powder

Pure copper powders from two different powder suppliers were used: deoxygenated oxygen-free pure copper (Cu-OF) and oxygenated electrolytic tough pitch copper (Cu-ETP). Both powders were gas atomized. The particle size distribution and morphology were verified using the CAMSIZER X2 (Microtrac Retsch GmbH, Haan, Germany) based on the dynamic digital image analysis according to ISO 13322-2. The chemical composition was provided by the material suppliers and hot carrier gas extraction using the inductar® ONH cube (Elementar Analysensysteme GmbH, Langenselbold, Germany). The absorptivity of the powders was measured using a UV-VIS Zeiss MCS400 (Carl Zeiss Spectroscopy GmbH, Jena, Germany) and FT-NIR spectrometer Bruker Vertex 70 (Bruker Corporation, Billerica, MA, USA). To verify the morphology and satellites, additional SEM imaging was used.

2.2. Machine

The samples were manufactured using a TruPrint 1000 Green Edition (TRUMPF GmbH + Co. KG (Holding), Ditzingen, Germany) with an integrated TruDisk1020 disk laser with a wavelength of 515 nm and a maximum laser power of 500 W. All specimens were processed with the same parameter set: line energy input 0.808 J/mm, hatch distance 120 μm, and a layer thickness of 30 μm.

2.3. Samples

Copper cubes with $10 \times 10 \times 15$ mm^3 and vertical walls of 20×20 mm^2 with different thicknesses of 500 μm, 1 mm, 1.5 mm, and 3 mm were built. Cylinders 8 mm diameter × 43 mm height for static mechanical testing were built in the vertical, horizontal, and 45° building direction (see Figure 1) and then machined to the geometry DIN 50125 B 4 × 20.

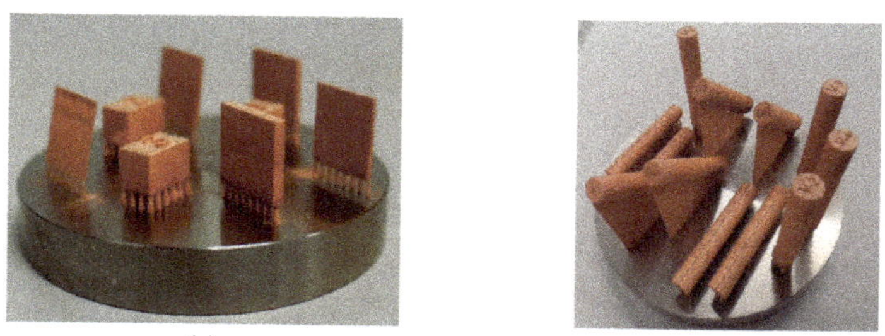

(**a**) (**b**)

Figure 1. Steel build plate 100 mm in diameter with (**a**) density cubes $10 \times 15 \times 10$ mm^3 and vertical walls 20×20 mm^2 of different thicknesses and (**b**) cylinders with 8 mm diameter × 43 mm height for tensile samples in three building directions.

To analyze the dimensional accuracy for certain features such as cylinders, overhangs, and walls, a benchmark geometry developed by the Fraunhofer Institute for Material and Beam Technology IWS within the AGENT-3D program was built and analyzed via 3D scanning using an ATOS Core GOM 135 (GOM GmbH, Braunschweig, Germany) which can be seen in Figure 2. The geometry and previous measurements with 3D scanning and computed tomography (CT) were described in detail by Lopez et al. [24] and Gruber et al. [25].

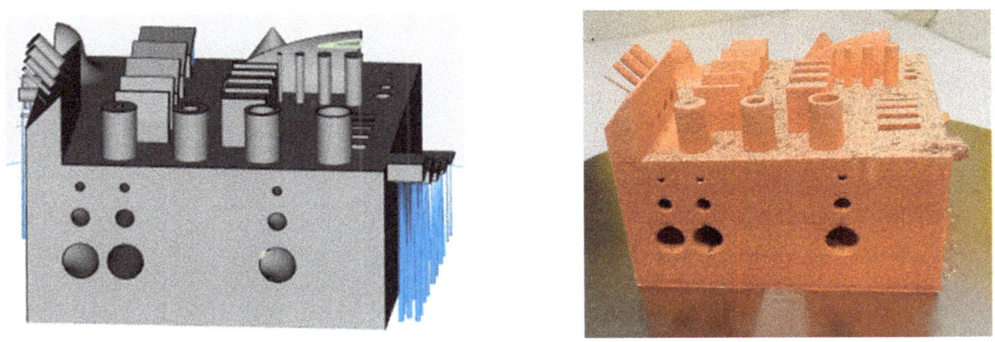

(**a**) (**b**)

Figure 2. Benchmark geometry (**a**) build job preparation file with support structures (blue) and (**b**) pure copper structure built with Cu-OFHC.

2.4. Sample Preparation

After the LPBF process, the samples were separated from the steel substrate via electrical discharge machining. Each cube was microsectioned in two planes (x–y and x–z, see Figure 3) and prepared for optical porosity analysis via image analysis. The resulting porosity and mean pore sizes were averaged from a total of six microsections per powder and etched with Adler 10:1 for the analysis of the microstructure.

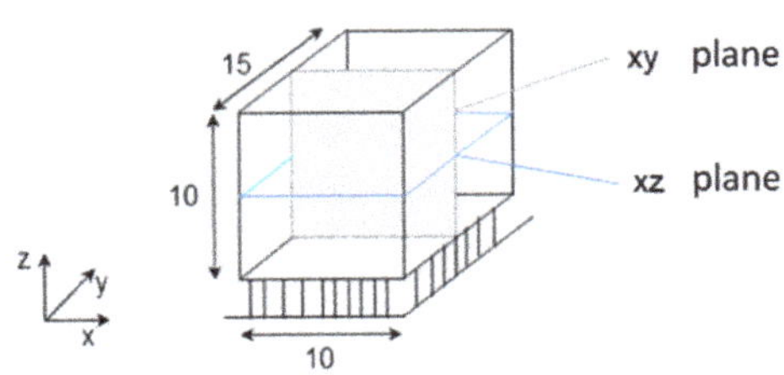

Figure 3. Definition of microsectioning planes.

The tensile cylinders were machined to B4x20 according to DIN EN ISO 50125 and tested with an inspect table 50 kN (Hegewald & Peschke Meß- und Prüftechnik GmbH, Nossen, Germany) according to DIN EN ISO 6892-1.

2.5. Conductivity Measurement

To measure the electrical conductivity, the eddy-current-based device SigmaScope 350 (Helmut Fischer GmbH, Sindelfingen, Germany) was used with an examination frequency of 120 kHz for the microsectioned cubes (x–z plane) and vertical walls (x–y plane) under the two surface conditions sand-blasted and milled. According to ASTM E1004—17 [26], the sample thickness should be at least 2.6 δ where δ is the standard depth of penetration of the induced eddy-currents calculated by Equation (1)

$$\delta = \frac{660}{\sqrt{f \times \sigma}} (\text{mm}) \tag{1}$$

where

f = examination frequency in Hz, dependent on the sensor,
σ = electrical conductivity of the sample in IACS percentage

Each measurement was repeated six times on the same sample and averaged. The precision of the measurement was 0.09 MS/m for 120 kHz and the correctness less than 1% of the measured value according to Helmut Fischer GmbH. With an expected 100% IACS and examination frequency of 120 kHz, the standard depth of penetration was 190 µm. Following the rule of a minimum wall thickness of 2.6 δ required a minimum sample thickness of 500 µm.

3. Results and Discussion

3.1. Powder Analysis

3.1.1. Morphology

The particle size distribution of both powders differed greatly as can be seen in Figure 4 and Table 1. D10 of Cu-OFHC was almost equal to D90 of Cu-ETP. Cu-ETP showed a bimodal behavior with peaks at 26 µm and 32 µm, whereas Cu-OFHC had a single peak at around 47 µm.

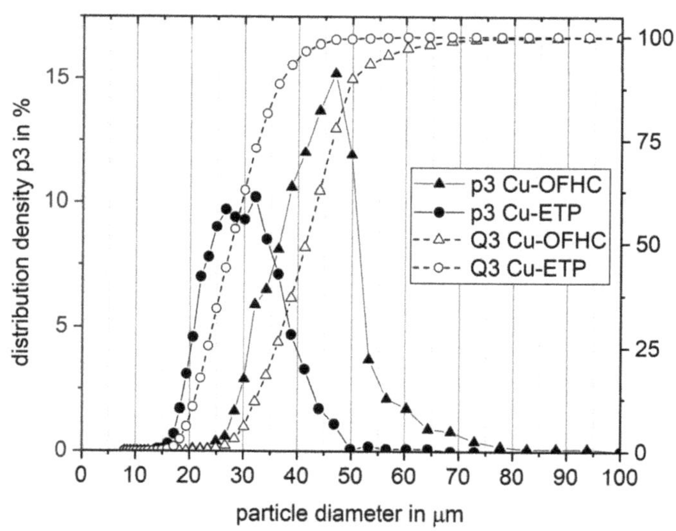

Figure 4. Particle size distribution of Cu-OFHC and Cu-ETP measured with the CAMSIZER X2.

Table 1. Main characteristics of particle size distribution of Cu-OFHC and Cu-ETP.

Powder	D10 in µm	D50 in µm	D90 in µm	Sphericity
Cu-OFHC	31.6	41.5	50.1	0.919
Cu-ETP	19.5	26.2	34.9	0.923

SEM imaging of both powders confirmed the presence of larger pores (see Figure 5) and the high sphericity values from the dynamic imaging showed a similar spherical shape and low amount of agglomerated particles. Additionally, no internal pores were observed for both powders from the embedded microsections.

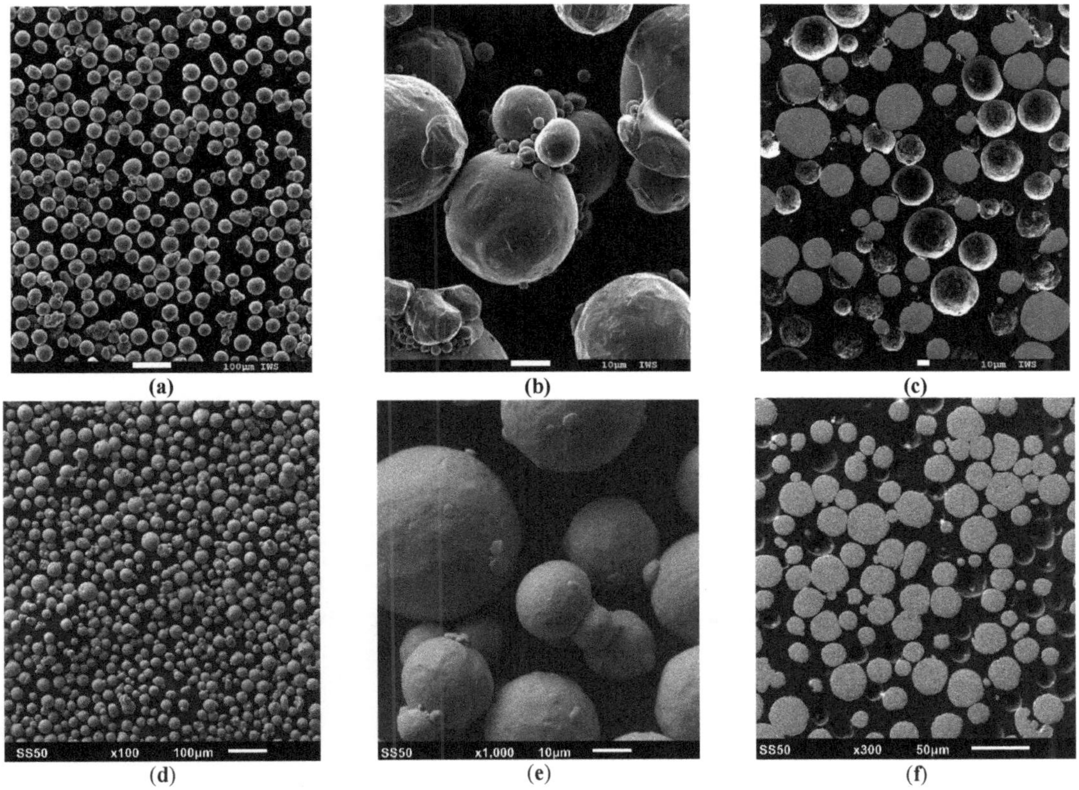

Figure 5. SEM images of loose Cu-OFHC powder at (**a**) 100× magnification, (**b**) 1000× magnification, and (**c**) embedded 300× magnification; SEM images of loose Cu-ETP powder at (**d**) 100× magnification, (**e**) 1000× magnification, and (**f**) embedded 300× magnification.

3.1.2. Chemical Composition

Despite the different grades provided, both powders had a similar chemical composition according to supplier specifications and hot carrier gas extraction measurements (see Table 2). One explanation for this could be the oxygen uptake of the Cu-OFHC powder during powder production and powder handling prior to processing, leading to similar results compared to the Cu-ETP powder.

Table 2. Cu content provided by the suppliers and H, N, O content of Cu-OFHC and Cu-ETP powder measured with hot carrier gas extraction.

	Supplier Information	Hot Gas Extraction			Technical Specification DIN CEN/TS 13388
	Cu	H	N	O	O
	in wt. %	in ppm			in %
Cu-OFHC	99.95	10	100	230	[1]
Cu-ETP	99.97	10	30	270	0.04–0.06

[1] Oxygen content must ensure hydrogen resistance according to DIN EN 1976.

3.1.3. Absorptivity

The absorptivity A of both pure copper powders at 515 and 1064 nm shown in Table 3 emphasize the advantages of using a green laser source, since the absorption increases by 260% compared to using a conventional infrared laser source.

Table 3. Absorptivity A at 515 nm and 1064 nm measured with UV-VIS spectrometry.

Powder	A at 515 nm	A at 1064 nm
Cu-OFHC	72.21%	27.27%
Cu-ETP	76.93%	31.99%

Cu-ETP showed higher absorption for both wavelengths, which could be explained by the smaller particle size resulting in larger surface area and increased multiple scattering of the laser beam in the powder bed that was also observed by Gu et al. [27]. Overall, both powders had similar morphological and chemical properties and the main difference was the particle size distribution.

3.2. Porosity

Light microscopy of the two microsectioned planes of the density cubes revealed a dense core and porous surface for both powders (see Figure 6) with a core porosity of 0.013% for the Cu-ETP powder and 0.017% for the Cu-OFHC powder (see Table 4).

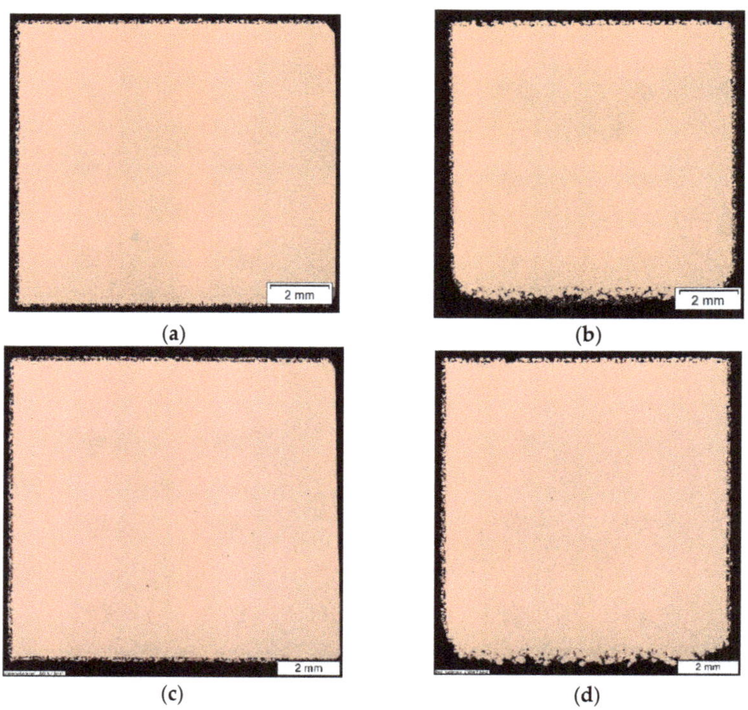

Figure 6. Polished microsection of Cu-OFHC in the (**a**) x–y plane and (**b**) x–z plane and polished microsection of Cu-ETP in the (**c**) x–y plane and (**d**) x–z plane.

Table 4. Porosity analysis of three density cubes per powder via image analysis.

Powder	Mean Porosity in %	Mean Max. Pore Size in μm	Mean Porous Layer Thickness in μm
Cu-OFHC	0.017 ± 0.0243	410.0 ± 8.50	152 ± 17.1
Cu-ETP	0.013 ± 0.0094	52.5 ± 27.59	144 ± 35.5

Again, the Cu-ETP exhibited slightly higher density values, which could be attributed to the smaller particle size resulting in higher packing and higher absorption. At the same time, it should be noted that microsections only reveal the behavior in one plane and the measured difference in density could also be statistically insignificant.

3.3. Microstructure

The microstructure in the x–z plane for both powders showed the typical grain growth parallel to the building direction (see Figure 7). Even though the measured porosity was very low, single spots of lack of fusion could be detected in both samples. The scanning tracks were also faintly visible and gave insight on the track width and penetration depth of the laser. The minimum track width was mainly restricted by the used 200 μm laser spot size. Density cubes from both powders showed similar microstructural evolution.

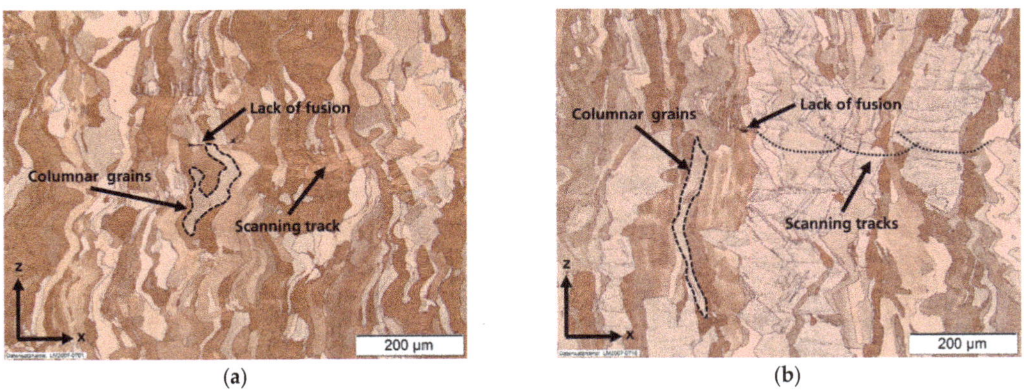

Figure 7. Microstructure of (**a**) Cu-OFHC in the x–z plane and (**b**) Cu-ETP in the x–z plane.

3.4. Electrical Conductivity

The electrical conductivity was measured from the prepared x–z microsections of the previously analyzed three density cubes for each powder (see Table 5).

Table 5. Comparison of electrical conductivity measured with the eddy-current method on polished microsectioned surfaces.

	MS/m	% IACS
Cu-OFHC	58.12 ± 0.26	100.0 ± 0.44
Cu-ETP	57.34 ± 0.26	98.6 ± 0.44

The slight difference of measured electrical conductivity using three cubes built from both powders (0.7 MS/m or 1.4% IACS) could be insignificant. This should be investigated in a later study with more samples. The electrical conductivity measurements were continued for Cu-OFHC vertical walls of different wall thicknesses only, since the density measurements, microstructure, and bulk electrical conductivity for both powders were very similar. A large discrepancy of the electrical conductivity was observed between a sand-blasted surface and a milled surface (see Figure 8) and the maximum values from the bulk conductivity could not be reproduced. This effect of surface condition increased with smaller wall thicknesses. At a 3 mm wall thickness, the difference was 8.4% and at 1 mm wall thickness the discrepancy increased to 34.6% while the standard deviation also increased. It is thought that perhaps the remaining air in the porous layer (see Figure 6) acted as an isolator, which then reduced the measured conductivity. This could explain the lower conductivity measurements of the 500 μm wall.

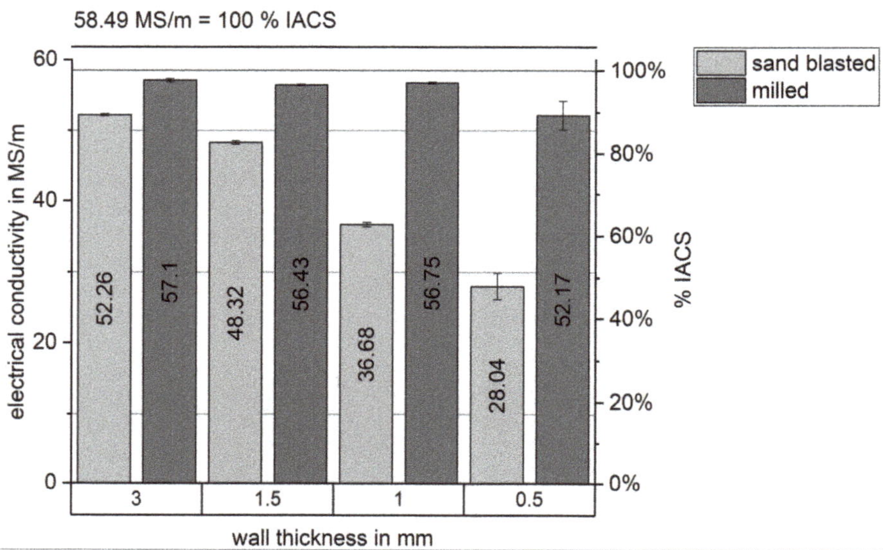

Figure 8. Effect of surface condition and wall thickness on measured electrical conductivity of Cu-OFHC.

3.5. Static Mechanical Properties

The mechanical properties (see Table 6) were obtained only from the Cu-OFHC powder due to the low differences in density, microstructure, and electrical conductivity of the cubes from both powders. Significant anisotropic behavior was found in the three building directions. Vertically and horizontally built samples showed similar behavior, whereas the diagonal samples showed a lower Young's modulus and lower strength. The mechanical properties of the additively manufactured copper were in the range of conventional soft annealed pure copper, according to [28] (see Table 6).

Table 6. Tensile testing results of Cu-OFHC.

Condition (Building Direction or Conventional)	Young's Modulus E in GPa	Yield Strength $R_{p0.2}$ in MPa	Ultimate Tensile Strength R_m in MPa	Elongation at Break A in %
Vertical	130.6 ± 27.6	135.7 ± 2.3	212.3 ± 3.8	51.5 ± 8.4
Diagonal	90.0 ± 9.6	127.3 ± 2.1	187.7 ± 2.1	47.0 ± 3.2
Horizontal	144.3 ± 15	134.8 ± 2.5	224.3 ± 2.2	47.4 ± 3.5
Soft annealed Cu-OF acc. to [28]	110	<100	200–250	40–60

3.6. Geometrical Accuracy

Features smaller than 0.5 mm are not converted into scanning vectors in the slicing software Materialise Trumpf Build Processor and Magics 24.1 due to the set-up beam compensation of 100 μm in combination with the beam diameter of 200 μm. Therefore, the smallest features of the benchmark geometry with thicknesses of 100 μm were not scanned and could therefore not be analyzed. The geometrical accuracy was highest for vertical cylinders larger or equal to 1 mm in diameter (see Figure 9). The increased relative deviation at a diameter of 500 µm is attributed to the large laser focus and beam compensation. For vertical walls, the deviation was feature size independent around 3 to 4%. Overall, features larger than 500 µm had a very high accuracy.

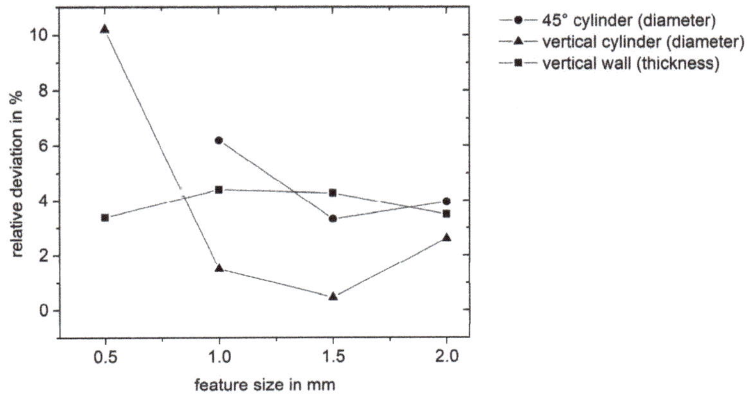

Figure 9. Relative deviation from the feature size of three different geometrical features (vertical wall, vertical cylinder, and 45° cylinder) analyzed from 3D scanning data points from the Cu-OFHC benchmark geometry.

All cooling channels 1, 2, and 4 mm in diameter (straight and curved) could be built without support structures and the remaining powder could be removed with pressurized gas (see Figure 2b). This showed the potential for complex inner cooling channels in future pure copper applications using LPBF having an advantage compared to electron beam melting, where unused powder is agglomerated in a sinter cake, making the removal of powder in inner channels difficult to almost impossible.

4. Conclusions

The following conclusions could be made in this study:

- Density above 99.8% and bulk electrical conductivity values of 98.6% IACS and 100% IACS for the two pure copper powders with different particle size distributions were achieved in accordance with [23], proving that the integration of the green laser source into the TruPrint1000 results in a stable process to build high-quality pure copper samples.
- A ground or polished surface with an area of 15 × 15 mm^2 and a sample thickness of at least 1.5 mm results in a reproducible and correct measurement of the electrical conductivity with the eddy-current method.
- The mechanical properties show anisotropic behavior, which was expected due to the layer-wise build up. The highest strength was found in horizontally build tensile samples.
- Features smaller than 500 µm are difficult to achieve due to the laser focus diameter of 200 µm, the scanning strategy of using contour lines for every feature, and the processing software.

Future work will focus on the influence of oxygen pick up within the powder during the build process and during powder storage on the part quality and powder recyclability as well as geometrical capabilities with the 200 µm laser focus diameter and minimum powder requirements for high-density and high-conductivity parts.

Author Contributions: Conceptualization, methodology, software, validation, investigation, writing—original draft preparation, visualization: S.G., F.B., and C.L.; Conceptualization, writing—review and editing: L.S., E.L., F.B., and C.L.; supervision: L.S.; project administration: E.L. All authors have read and agreed to the published version of the manuscript.

Funding: These experiments were performed as part of the Fraunhofer High-Performance Center "Smart Production and Materials" program.

Institutional Review Board Statement: Not Applicable.

Informed Consent Statement: Not Applicable.

Data Availability Statement: The data can be requested from the corresponding author.

Conflicts of Interest: There were no conflicts of interest. The funders had no role in the design of the study; in the collection, analyses, or interpretation of data; in the writing of the manuscript; or in the decision to publish the results.

References

1. Lomakin, K.; Guschlbauer, R.; Osmanlic, F.; Fu, Z.; Sippel, M.; Helmreich, K.; Korner, C.; Vossiek, M.; Gold, G. 3D Printed Copper Waveguides by Selective Electron Beam Melting Process for E-Band. In Proceedings of the 49th European Microwave Conference (EuMC), Paris, France, 1–3 October 2019.
2. Guschlbauer, R.; Burkhardt, A.K.; Fu, Z.; Körner, C. Effect of the oxygen content of pure copper powder on selective electron beam melting. *Mater. Sci. Eng. A* **2020**, *779*, 139106. [CrossRef]
3. Lodes, M.A.; Guschlbauer, R.; Körner, C. Process development for the manufacturing of 99.94% pure copper via selective electron beam melting. *Mater. Lett.* **2015**, *143*, 298–301. [CrossRef]
4. Raab, S.J.; Guschlbauer, R.; Lodes, M.A.; Körner, C. Thermal and Electrical Conductivity of 99.9% Pure Copper Processed via Selective Electron Beam Melting. *Adv. Eng. Mater.* **2016**, *18*, 1661–1666. [CrossRef]
5. Bai, Y.; Williams, C. An exploration of binder jetting of copper. *Rapid Prototyp. J.* **2015**, *21*, 177–185. [CrossRef]
6. Yan, X.; Chang, C.; Dong, D.; Gao, S.; Ma, W.; Liu, M.; Liao, H.; Yin, S. Microstructure and mechanical properties of pure copper manufactured by selective laser melting. *Mater. Sci. Eng. A* **2020**, *789*, 139615. [CrossRef]
7. Silbernagel, C.; Gargalis, L.; Ashcroft, I.; Hague, R.; Galea, M.; Dickens, P. Electrical resistivity of pure copper processed by medium-powered laser powder bed fusion additive manufacturing for use in electromagnetic applications. *Addit. Manuf.* **2019**, *29*, 100831. [CrossRef]
8. Kaden, L.; Matthäus, G.; Ullsperger, T.; Engelhardt, H.; Rettenmayr, M.; Tünnermann, A.; Nolte, S. Selective laser melting of copper using ultrashort laser pulses. *Appl. Phys. A* **2017**, *123*, 596. [CrossRef]
9. Jadhav, S.; Dadbakhsh, S.; Goossens, L.; Kruth, J.-P.; Van Humbeeck, J.; Vanmeensel, K. Influence of selective laser melting process parameters on texture evolution in pure copper. *J. Mater. Process. Technol.* **2019**, *270*, 47–58. [CrossRef]
10. Colopi, M.; Caprio, L.; Demir, A.; Previtali, B. Selective laser melting of pure Cu with a 1 kW single mode fiber laser. *Procedia CIRP* **2018**, *74*, 59–63. [CrossRef]
11. Becker, D. Selektives Laserschmelzen von Kupfer und Kupferlegierungen. Master Dissertation, RWTH Aachen University, Aachen, Germany, 2014.
12. Colopi, M.; Demir, A.G.; Caprio, L.; Previtali, B. Limits and solutions in processing pure Cu via selective laser melting using a high-power single-mode fiber laser. *Int. J. Adv. Manuf. Technol.* **2019**, *104*, 2473–2486. [CrossRef]
13. Lingqin, X.; Guang, C.; Luyu, Z.; Pan, L. Explore the feasibility of fabricating pure copper parts with low-laser energy by selective laser melting. *Mater. Res. Express* **2020**, *7*, 106509. [CrossRef]
14. Lykov, P.; Safonov, E.; Akhmedianov, A. Selective Laser Melting of Copper. *Mater. Sci. Forum* **2016**, *843*, 284–288. [CrossRef]
15. Zhang, X.; Sun, C.; Pan, T.; Flood, A.; Zhang, Y.; Li, L.; Liou, F. Additive manufacturing of copper–H13 tool steel bi-metallic structures via Ni-based multi-interlayer. *Addit. Manuf.* **2020**, *36*, 101474. [CrossRef]
16. Prasad, H.S.; Brueckner, F.; Volpp, J.; Kaplan, A.F.H. Laser metal deposition of copper on diverse metals using green laser sources. *Int. J. Adv. Manuf. Technol.* **2020**, *107*, 1559–1568. [CrossRef]
17. Asano, K.; Tsukamoto, M.; Sechi, Y.; Sato, Y.; Masuno, S.-I.; Higashino, R.; Hara, T.; Sengoku, M.; Yoshida, M. Laser metal deposition of pure copper on stainless steel with blue and IR diode lasers. *Opt. Laser Technol.* **2018**, *107*, 291–296. [CrossRef]
18. Boley, C.D.; Khairallah, S.A.; Rubenchik, A.M. Calculation of laser absorption by metal powders in additive manufacturing. *Appl. Opt.* **2015**, *54*, 2477–2482. [CrossRef] [PubMed]
19. El-Wardany, T.; She, Y.; Jagdale, V.; Garofano, J.K.; Liou, J.; Schmidt, W. Challenges in 3D Printing of High Conductivity Copper. In Proceedings of the ASME 2017 International Technical Conference and Exhibition on Packaging and Integration of Electronic and Photonic Microsystems, San Francisco, CA, USA, 29 August–1 September 2017.
20. Tran, T.Q.; Chinnappan, A.; Lee, J.K.Y.; Loc, N.H.; Tran, L.T.; Wang, G.; Kumar, V.V.; Jayathilaka, W.A.D.M.; Ji, D.; Doddamani, M.; et al. 3D Printing of Highly Pure Copper. *Metals* **2019**, *9*, 756. [CrossRef]

21. Trevisan, F.; Calignano, F.; Lorusso, M.; Lombardi, M.; Manfredi, D.; Fino, P. Selective laser melting of chemical pure copper powders. In Proceedings of the Euro PM2017 Congress & Exhibition, Milan, Italy, 1–5 October 2017.
22. TIkeshoji, T.-T.; Nakamura, K.; Yonehara, M.; Imai, K.; Kyogoku, H. Selective Laser Melting of Pure Copper. *JOM* **2018**, *70*, 396–400. [CrossRef]
23. Wagenblast, P.; Myrell, A.; Thielmann, M.; Scherbaum, T.; Coupek, D. Additive manufacturing with green disk lasers. In *Laser 3D Manufacturing VII, Proceedings of the Society of Photo-Optical Instrumentation Engineers (SPIE) Conference Series, San Francisco, CA, USA, 1–6 February 2020*; Gu, B., Chen, H., Helvajian, H., Eds.; SPIE—International Society for Optics and Photonics: Bellingham, WA, USA, 2020; p. 18. [CrossRef]
24. Lopez, E.; Felgueiras, T.; Grunert, C.; Brückner, F.; Riede, M.; Seidel, A.; Marquardt, A.; Leyens, C.; Beyer, E. Evaluation of 3D-printed parts by means of high-performance computer tomography. *J. Laser Appl.* **2018**, *30*, 032307. [CrossRef]
25. Gruber, S.; Grunert, C.; Riede, M.; López, E.; Marquardt, A.; Brueckner, F.; Leyens, C. Comparison of dimensional accuracy and tolerances of powder bed based and nozzle based additive manufacturing processes. *J. Laser Appl.* **2020**, *32*, 032016. [CrossRef]
26. ASTM E1004-17. *Test Method for Determining Electrical Conductivity Using the Electromagnetic (Eddy-Current) Method*; ASTM International: West Conshohocken, PA, USA, 2017.
27. Gu, D.; Yang, Y.; Xi, L.; Yang, J.; Xia, M. Laser absorption behavior of randomly packed powder-bed during selective laser melting of SiC and TiB2 reinforced Al matrix composites. *Opt. Laser Technol.* **2019**, *119*, 105600. [CrossRef]
28. DIN EN 13599:2014-12. *Kupfer und Kupferlegierungen-Platten, Bleche und Bänder aus Kupfer für die Anwendung in der Elektrotechnik*; Deutsche Fassung EN_13599:2014; VDE Verband der Elektrotechnik Elektronik Informationstechnik e.V.: Frankfurt am Main, Germany, 2014.

materials

Article

Microstructural Variations in Laser Powder Bed Fused Al–15%Fe Alloy at Intermediate Temperatures

Wenyuan Wang [1,*], Naoki Takata [1,*], Asuka Suzuki [1], Makoto Kobashi [1] and Masaki Kato [2]

1 Department of Materials Process Engineering, Graduate School of Engineering, Nagoya University, Furo-cho, Chikusa-ku, Nagoya 464-8603, Japan; suzuki.asuka@material.nagoya-u.ac.jp (A.S.); kobashi.makoto@material.nagoya-u.ac.jp (M.K.)
2 Aichi Center for Industry and Science Technology, 1267-1 Akiai, Yakusa-cho, Toyota 470-0356, Japan; masaki_2_katou@pref.aichi.lg.jp
* Correspondence: wang.wenyuan.m4@s.mail.nagoya-u.ac.jp (W.W.); takata.naoki@material.nagoya-u.ac.jp (N.T.)

Abstract: The samples of the Al–15Fe (mass%) binary alloy that were additively manufactured by laser powder bed fusion (L-PBF) were exposed to intermediate temperatures (300 and 500 °C), and the thermally induced variations in their microstructural characteristics were investigated. The L-PBF-manufactured sample was found to have a microstructure comprising a stable θ-$Al_{13}Fe_4$ phase localized around melt-pool boundaries and several spherical metastable Al_6Fe-phase particles surrounded by a nanoscale α-Al/Al_6Fe cellular structure in the melt pools. The morphology of the θ phase remained almost unchanged even after 1000 h of exposure at 300 °C. Moreover, the nanoscale α-Al/Al_6Fe cellular structure dissolved in the α-Al matrix; this was followed by the growth (and nucleation) of the spherical Al_6Fe-phase particles and the precipitation of the θ phase. Numerous equiaxed grains were formed in the α-Al matrix during the thermal exposure, which led to the formation of a relatively homogenous microstructure. The variations in these microstructural characteristics were more pronounced at the higher investigated temperature of 500 °C.

Keywords: additive manufacturing; aluminum alloys; microstructure; intermetallics; thermal exposure

Citation: Wang, W.; Takata, N.; Suzuki, A.; Kobashi, M.; Kato, M. Microstructural Variations in Laser Powder Bed Fused Al–15%Fe Alloy at Intermediate Temperatures. *Materials* **2022**, *15*, 4497. https://doi.org/10.3390/ma15134497

Academic Editors: Manfred Schmid, Bilal Gökce and Eric Jägle

Received: 30 May 2022
Accepted: 23 June 2022
Published: 26 June 2022

1. Introduction

Metal additive manufacturing (AM) is an advanced manufacturing technology used for fabricating complex-shaped metal/alloy components using computer-aided design [1]. Laser powder bed fusion (L-PBF) is one of the representative metal-AM processes [2] that has been adopted for a considerably lightweight aluminum (Al) alloy series [3,4]. In L-PBF, a scanning laser irradiates the metal/alloy powder that is sequentially bedded on a base plate, which leads to selective melting and subsequent rapid solidification. Consequently, L-PBF-processed Al alloy products exhibit significantly refined microstructures [5–8], which lead to higher strengths than those of the Al alloys fabricated using conventional casting processes [9–12]. Al–Si-based alloys are known to be suitable for the L-PBF process [13–15]; however, L-PBF-processed Al–Si-based alloys exhibit reduced strength at temperatures higher than 200 °C [16]. The potential application of L-PBF-manufactured Al alloys in radial impellers operating at intermediate temperatures above 200 °C (inside the vehicle turbochargers) has encouraged the development of new Al alloys with superior strength at both ambient and intermediate temperatures. To accommodate the demand for materials with high-temperature strength, a variety of heat-resistant Al alloys, such as Al–Cr, Al–Mn, Al–Ni, Al–Ni–Fe, and Al–Ce–Mn, have been proposed for fabrication by L-PBF [17–19].

With the aim of fabricating Al alloys using common alloy elements instead of rare-earth elements, attempts have been made to investigate the feasibility of adopting L-PBF to fabricate an Al–Fe binary alloy with a high Fe content (15 mass%) [19], which corresponds to a hyper-eutectic composition in the Al–Fe binary system. In general, coarsened Al–Fe intermetallic compounds (stable $Al_{13}Fe_4$ phase) were often formed in the cast Al–Fe-based alloys with high Fe contents. The brittle Al-rich intermetallic phases have a detrimental effect on the ductility of the materials. However, the L-PBF-manufactured Al–15%Fe alloy exhibits refined microstructures [20,21] containing numerous nanosized particles of the metastable Al_6Fe phase [22]. Moreover, the L-PBF-manufactured Al–15%Fe alloy shows a high yield strength of about 400 MPa at 300 °C [23], which is higher than that of both the 8xxx alloy series [24,25] (Al–Fe-based alloys used in powder metallurgy) and the L-PBF-manufactured Al-based multi-element alloys [18,26]. The hardness of these specimens slightly decreases after long-term thermal exposure, suggesting that the high thermal stability of the nanosized metastable Al_6Fe phase strengthens the L-PBF-manufactured Al–Fe alloys. This indicates the remarkable potential of Al–15%Fe as a lightweight Al alloy that can be additively manufactured for high-temperature applications. However, the metastable Al_6Fe phase can transform into a stable $Al_{13}Fe_4$ phase [27] that is in equilibrium with the α-Al matrix after long-term exposure to high temperatures. Additionally, the changes in the microstructural features of L-PBF-processed Al–Fe binary alloys during thermal exposure are not fully understood.

Therefore, the microstructures of the L-PBF-processed Al–15%Fe alloy exposed to intermediate temperatures (300 and 500 °C), which contained refined Al–Fe intermetallic phases, were systematically characterized in this study to elucidate the microstructural variations and the phase transformation from metastable Al_6Fe to stable $Al_{13}Fe_4$.

2. Materials and Methods

An Al–15Fe (mass%) binary alloy powder with an average particle size below 30 μm was prepared via gas atomization; the details concerning the preparation of the alloy powder can be found elsewhere [20]. Rectangular samples with the approximate dimensions of $15 \times 15 \times {\sim}5$ mm^3 were constructed using a ProX DMP 200 machine (3D Systems, Rock Hill, SC, USA). The following optimized L-PBF parameters were used to manufacture the samples [14]: laser scanning speed, 0.4 m/s; laser power, 128 W; hatch distance, 0.1 mm; powder bed layer thickness, 0.03 mm; and beam focus size, ~0.1 mm. The scanning laser patterns were separated in each 10 mm-sized hexagonal grid, and the direction of the laser scanning was consecutively altered by 90° for each powder layer [28]. The constructed samples had high relative densities (>96%). The as-built samples were exposed to 300 and 500 °C for various periods ranging from 1 h to 1000 h, followed by a water quench.

The samples exposed to high temperatures for different durations were embedded in resin and then mechanically polished with SiC paper. Scanning electron microscopy (SEM; JSM-IT500 and JSM-6610A, JEOL Ltd., Tokyo, Japan) was performed to examine the microstructures of the prepared samples. To facilitate these observations, the sample surfaces were polished with 0.05 μm-sized colloidal silica particles (pH 9.8 in liquid). Vickers hardness tests were performed on these samples using a test load of 1.98 N and a loading duration of 15 s at room temperature. X-ray diffraction (XRD) measurements were carried out using a Rigaku ULTIMA IV instrument equipped with a Cu radiation source at 40 kV. The cross-sectional samples were subjected to argon-ion polishing using a cross-section polisher at 6 V. Orientation analyses were performed using the electron backscatter diffraction (EBSD) technique with scanning step sizes of 0.2 or 0.3 μm. Thin samples were prepared from the thermally exposed specimens for transmission electron microscopy (TEM) analysis. The pieces were cut into a plate shape using a low-speed cutter and then polished with SiC paper to prepare foil samples with a thickness of ~0.1 mm. The thin-foil samples were subjected to argon-ion polishing at 6.0 kV using an Ion Slicer™ (JEOL EM-09100IS) and then smoothened for approximately 600 s at a low voltage (2.0 kV) for TEM analysis. The microstructures of the prepared samples were characterized by TEM (JEOL JEM-2100F/HK) at 200 kV.

3. Results and Discussion

3.1. Microstructure of As-Built Sample

Figure 1 shows the multi-scale microstructural characteristics of the as-built Al–15%Fe alloy sample. The low-magnification SEM images [Figure 1a,b] show representative melt-pool structures, which refer to the laser-scanning tracks in which the regions are selectively melted and rapidly solidified. The high-magnification SEM image (Figure 1c) shows many coarse particles of the Al–Fe intermetallic phase, with a leaf-shaped morphology localized along the melt-pool boundaries. Comprehensive EBSD analyses [21] have confirmed the formation of the θ-$Al_{13}Fe_4$ stable phase [27] that is in equilibrium with the α-Al matrix in the Al–Fe binary system. Numerous particles of the metastable Al_6Fe phase [22], several hundred nanometers in size, were distributed in the melt pools (Figure 1c,d). The TEM observation (Figure 1d) revealed that these spherical Al_6Fe-phase particles were surrounded by nanoscale cellular structures composed of α-Al and Al_6Fe phases in the melt-pool structure. The formation sequences of the nanosized metastable phase and the relatively coarse stable phase during solidification have been clarified [21] using the equilibrium and non-equilibrium phase diagrams of the Al–Fe system [29].

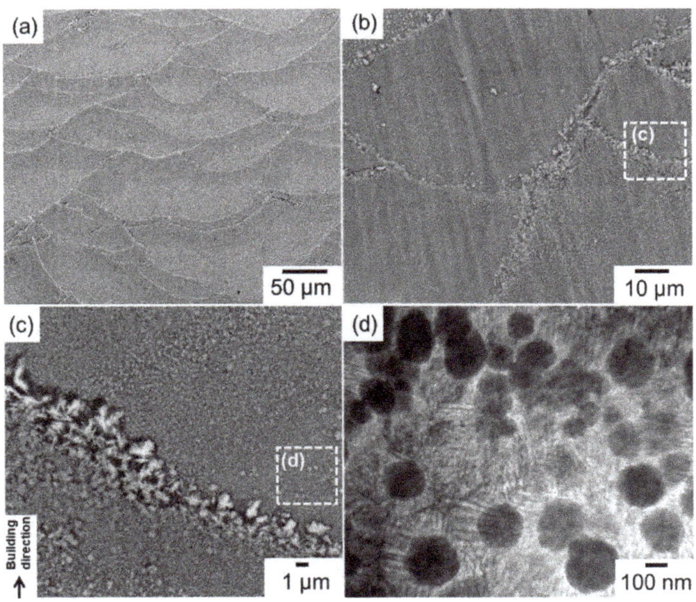

Figure 1. (**a–c**) SEM images showing microstructures of L-PBF-constructed Al–15%Fe alloy: (**a**,**b**) low-magnification views and (**c**) a location around the melt-pool boundary. (**d**) Bright-field TEM image showing spherical Al_6Fe-phase particles surrounded by nanoscale cellular structures in the melt pool.

3.2. Variation in Al–Fe Intermetallic Phases at Elevated Temperatures

Figure 2 shows representative XRD profiles of the as-built and thermally exposed Al–15%Fe alloy samples. The XRD profile of the as-built sample confirmed the presence of metastable Al_6Fe and stable θ-$Al_{13}Fe_4$ phases in the α-Al matrix, which is consistent with the microstructural characterization results shown in Figure 1. The samples exposed to 300 °C exhibited higher diffraction intensities from the Al_6Fe and θ phases; this tendency was enhanced after the long-term exposure for 1000 h. In contrast, the sample exposed to 500 °C for 100 h exhibited considerably lower diffraction intensities from the Al_6Fe phase and similar θ-phase intensities to those detected in the samples exposed to 300 °C. This variation in the XRD profiles can be attributed to the dissolution of the metastable Al_6Fe phase upon exposure to 500 °C.

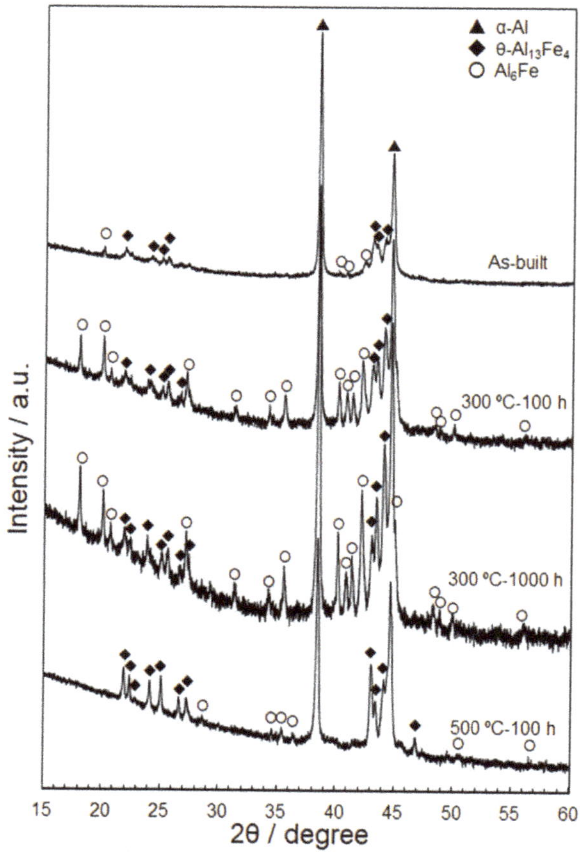

Figure 2. X-ray diffraction (XRD) profiles of the L-PBF-built pristine and thermally exposed Al–15%Fe alloy samples [300 °C (100 h, 1000 h) and 500 °C (100 h)].

Figure 3 shows the SEM-EBSD images that reveal the microstructures at different locations in the as-built and thermally treated samples (300 and 500 °C). The macroscopic melt-pool structure in the L-PBF-built sample changed minimally upon exposure to 300 °C, even after 1000 h [Figure 3a,c]. Moreover, the stable θ-$Al_{13}Fe_4$ phase that was localized along the melt-pool boundaries remained unchanged after 1000 h of exposure [Figure 3e,g]. Additionally, the size of the metastable Al_6Fe-phase particles located in the melt pools increased after 100 h of exposure [Figure 3i,j]; moreover, the number density increased considerably after 1000 h (Figure 3k). The larger volume of the Al_6Fe phase is consistent with the high diffraction intensities observed in the XRD profiles (Figure 2). These results demonstrate the nucleation and growth of the spherical Al_6Fe-phase particles after exposure to 300 °C. The quantitative analyses for the spherical Al_6Fe phase were described elsewhere [23]. However, the spherical Al_6Fe phase was scarcely found in the sample exposed to 500 °C for 100 h. Although macroscopic melt-pool structures were observed (Figure 3d), abundant coarse intermetallic-phase particles were present in the melt pools (Figure 3l) and at the melt-pool boundaries (Figure 3h). The initial leaf-shaped morphology of the θ phase transformed into a granular or plate-shaped morphology. These granular Al–Fe intermetallic phases were found in the melt pools.

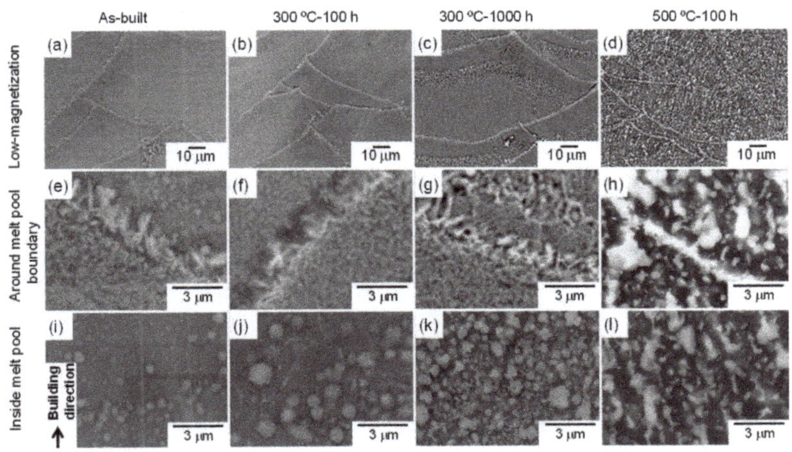

Figure 3. SEM images showing microstructures of (**a**,**e**,**i**) pristine and (**b**–**d**,**f**–**h**,**j**–**l**) thermally exposed L-PBF-built Al–15%Fe alloy samples: (**b**,**f**,**j**) 300 °C/100 h, (**c**,**g**,**k**) 300 °C/1000 h, and (**d**,**h**,**l**) 500 °C/100 h.

Figure 4 shows the bright-field TEM images that reveal the microstructures in the melt pools of the as-built and thermally treated samples (300 and 500 °C). The nanoscale cellular structure of the eutectic α-Al/Al_6Fe phases, which appeared in the as-built sample, was scarcely observed in the sample exposed to 300 °C for 100 h [Figure 4a,b]; however, various spherical Al_6Fe-phase particles, several hundred nanometers in size, were observed. After 1000 h of exposure, numerous plate-shaped precipitates and spherical Al_6Fe-phase particles were found in the α-Al matrix (Figure 4c). These fine precipitates are consistent with the SEM observations of the fine particles (Figure 3k). However, the plate-shaped precipitates were not observed after the exposure to 500 °C (Figure 4d). Moreover, many granular particles appeared to be connected to each other in the α-Al matrix.

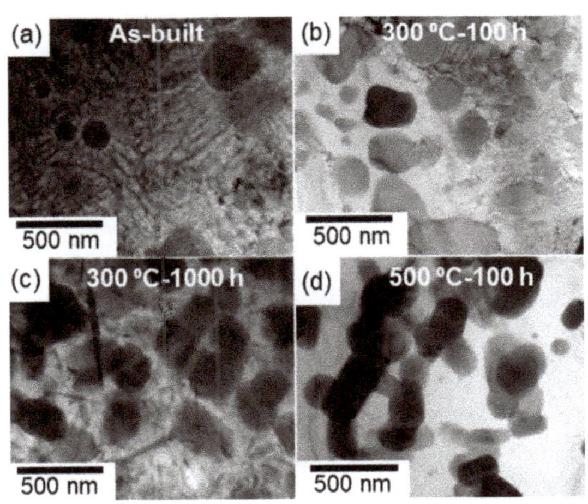

Figure 4. TEM images showing microstructures in the melt pools of the (**a**) as-built and (**b**–**d**) thermally treated L-PBF-built Al–15%Fe alloy samples: (**b**) 300 °C/100 h, (**c**) 300 °C/1000 h, and (**d**) 500 °C/100 h.

Selected area electron diffraction (SAED) patterns were subsequently acquired to identify the phases observed in the thermally exposed samples; the representative results are summarized in Figures 5 and 6. Each SAED pattern was captured from enclosed areas in the corresponding TEM images. For the coarse cellular structures that were locally observed in the sample exposed to 300 °C for 100 h (Figure 5a), the SAED pattern exhibited a ring diffraction configuration derived from the (222) plane of the Al_6Fe phase, indicating the presence of a relatively coarsened eutectic structure of the α-Al and Al_6Fe phases (Figure 5b). The SAED pattern of the sample exposed to 300 °C for 1000 h (Figure 5c), which was acquired from a relatively coarse precipitate with a plate-shaped morphology, as shown in Figure 4c, indicated that the incident beam was parallel to the [1–10] direction of the θ-$Al_{13}Fe_4$ stable phase with a monoclinic structure (Figure 5d). The observed morphology of the θ phase is consistent with that of the Al–2.5%Fe alloy manufactured by L-PBF [30]. The SAED pattern captured from the spherical particles that were several hundred nanometers in size (Figure 5e) revealed a clear diffraction pattern derived from the Al_6Fe phase (Figure 5f). The presence of the spherical Al_6Fe metastable phase is consistent with the XRD results (Figure 2). Numerous granular intermetallic phases, including several planar faults, were observed in the sample exposed to 500 °C for 100 h (Figure 6a). The corresponding SAED pattern displayed several diffractions derived from the θ phase, and the observed planar faults appeared to be parallel to the (001) plane of the θ phase (Figure 6b). These crystallographic features correspond well to the nanosized twins on the (001) plane in the grown θ phase [31,32]. These results clearly indicate the formation of a coarse stable θ phase in equilibrium with the α-Al phase. Additionally, the granular Al_6Fe phases remained localized, even after exposure to 500 °C for 100 h (Figure 6c,d).

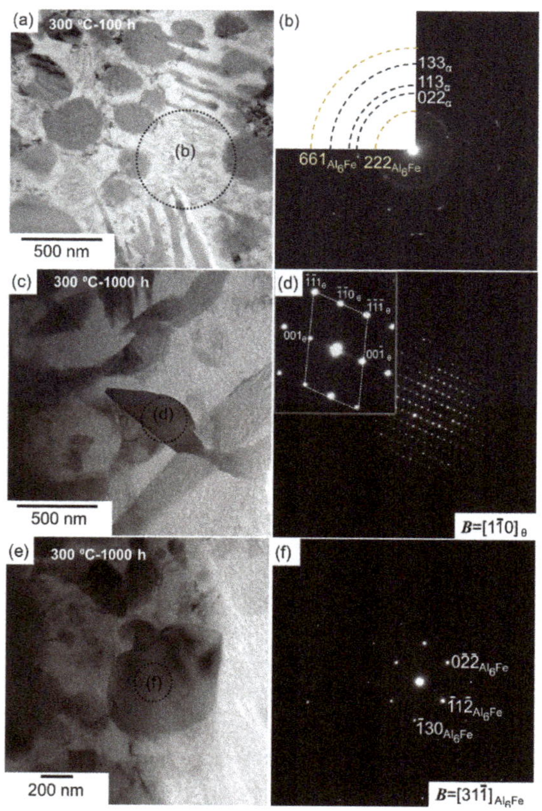

Figure 5. (**a**,**c**,**e**) Bright-field TEM images and (**b**,**d**,**f**) the corresponding SAED patterns of (**a**,**b**) pristine and (**c**–**f**) thermally exposed L-PBF-built samples (300 °C/1000 h).

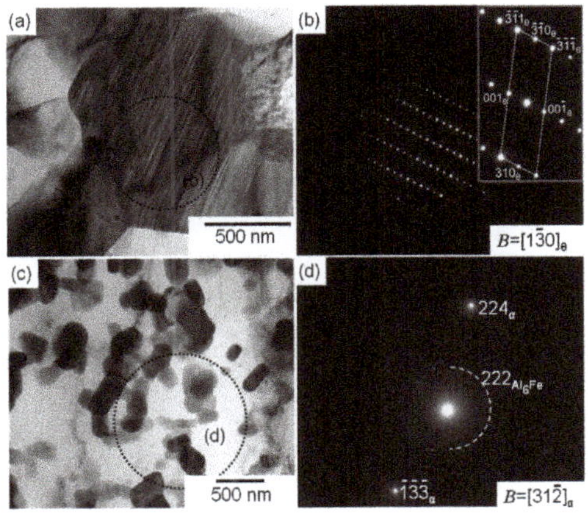

Figure 6. (**a**,**c**) Bright-field TEM images showing the intermetallic phases distributed in the melt pool and the corresponding (**b**,**d**) SAED patterns of an L-PBF-built Al–15%Fe alloy sample exposed to 500 °C for 100 h.

The aforementioned results indicate that in the samples exposed to 300 °C, the fine metastable Al_6Fe-phase particles in the eutectic cellular structure dissolve in the α-Al matrix; this is followed by the growth of the spherical metastable Al_6Fe-phase particles and the precipitation of the stable θ phase in the α-Al matrix. The growth of the stable θ phase is more pronounced at 500 °C. Moreover, the interface between the θ and Al_6Fe phases was scarcely observed in the samples exposed to the different temperatures, suggesting that the stable θ phase nucleated in the α-Al matrix containing the Fe solute, rather than at the interface of the metastable Al_6Fe phase with the α-Al matrix.

Figure 7 shows the changes in the Vickers hardness of the L-PBF-built Al–15%Fe alloy samples after the exposure to different temperatures. The as-built sample exhibited a high hardness of ~200 HV. The hardness decreased moderately with the increasing duration of the exposure to 300 °C. The slight reduction in hardness of these specimens is consistent with the gradual microstructural changes observed by SEM and TEM (Figures 3 and 4). However, the hardness of the specimen exposed to 500 °C significantly decreased to below 140 HV after 1 h and then continuously decreased to ~100 HV. This tendency is in good agreement with the significant change in microstructural features observed in Figures 3 and 4 through the replacement of the metastable Al_6Fe phase with the stable θ phase.

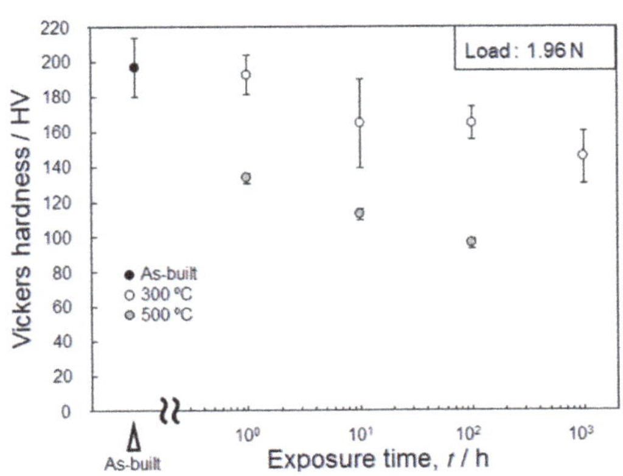

Figure 7. Changes in hardness of the L-PBF-built Al–15%Fe alloy samples as a function of exposure time at 300 and 500 °C.

3.3. Changes in α-Al Matrix at Elevated Temperatures

To examine the variations in the microstructural characteristics of the α-Al matrix in the specimens during their high-temperature exposure, the as-built and thermally exposed samples were subjected to an EBSD analysis. Figure 8 shows the orientation distribution maps of the fcc-structured α-Al matrix in the as-built and thermally treated specimens. In these maps, orientations parallel to the build direction are colored according to the attached unit triangle of the inverse pole figure. Several elongated grains with widths of a few tens of micrometers were predominantly formed in the as-built sample and were surrounded by high-angle grain boundaries with large misorientations (>15°); fine-grained microstructures (including unanalyzed areas) were also locally observed (Figure 8a). A comparison of the SEM-observed locations with the EBSD-analyzed areas revealed that the fine-grained regions were typically located inside the melt pools. The finely solidified Al_6Fe phases likely promoted nucleation during the solidification, resulting in local formation of the fine-grained α-Al phase. These microstructural morphologies remained almost unchanged even after 1000 h of exposure at 300 °C [Figure 8b,c]; nevertheless, relatively equiaxed grains were often observed. Additionally, the density of the low-angle boundaries (with misorientations smaller than 15°) decreased after the exposure. It is worth noting that a few equiaxed grains, several micrometers in size, were formed inside the local fine-grained regions (indicated by the arrowheads in Figure 8c), suggesting the occurrence of recrystallization at elevated temperatures. This trend was more evident in the sample exposed to 500 °C (Figure 8d), which resulted in the formation of a relatively homogenous microstructure in the α-Al matrix. The grain size ranged from approximately 1 to 10 μm. Grain boundary migration was suppressed by the pinning effect of the fine Al–Fe intermetallic phases dispersed in the α-Al matrix (Figure 3), which led to a relatively fine-grained microstructure even after exposure to 500 °C for 100 h (Figure 8d).

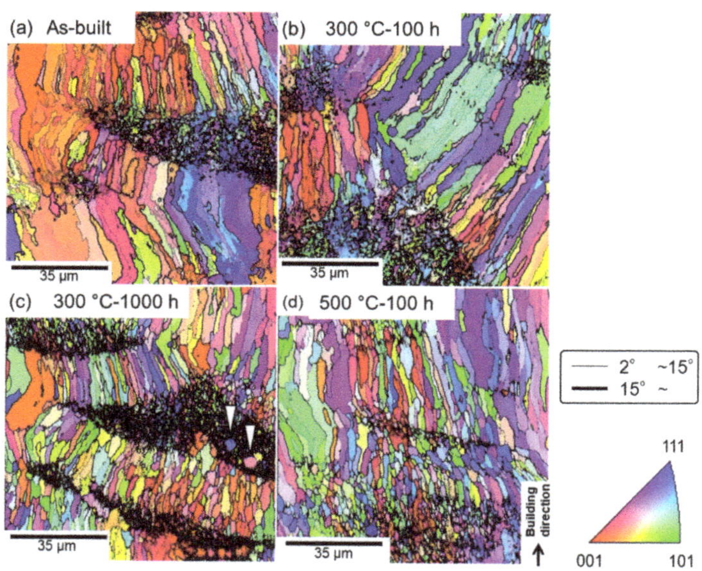

Figure 8. EBSD orientation maps of the fcc-structured α-Al matrix in the (**a**) as-built and (**b–d**) thermally exposed samples: (**b**) 300 °C/100 h, (**c**) 300 °C/1000 h, and (**d**) 500 °C/100 h. The fine lines correspond to misorientations (θ) of 2° < θ < 15°, whereas the bold lines represent θ > 15°.

Figure 9 shows the variations in the lattice parameter of the α-Al matrix in the thermally treated samples (300 and 500 °C) as a function of the exposure time. The lattice parameters were calculated using the XRD profiles shown in Figure 2. The scatted values of the measured lattice parameters suggest the inhomogeneous distribution of a solute Fe element in the α-Al matrix due to the complicated microstructures of the L-PBF-processed Al–15%Fe alloy samples. The lattice parameter of the as-built sample was ~0.405 nm, whereas that of the thermally treated specimens increased with the increasing exposure time and almost stabilized at ~0.406 nm. This trend was observed for both sets of the thermally exposed specimens (300 and 500 °C). The atomic radius of Fe (0.127 nm) is smaller than that of Al (0.143 nm) in the fcc structure [33]. Therefore, the observed increase in the lattice parameter could be due to the reduction in the Fe solute in the α-Al matrix upon exposure to elevated temperatures. The saturated value of the lattice parameter (~0.406 nm) is equivalent to that of the fully melted and slowly solidified samples of the used Al–15%Fe alloy powder [21] that solidifies at a low cooling rate of ~0.3 °C/s [34]. These results suggest that the Fe solute content almost achieved an equilibrium state after prolonged thermal exposure. Intriguingly, all the measured lattice parameters of the used Al–15%Fe alloy samples were greater than that of pure Al (0.40493 nm). The large lattice parameters of the α-Al phase in the L-PBF-built Al–15%Fe samples were presumably due to the presence of O solute atoms positioned at interstitial sites in the fcc structure. Composition analyses [20,21] have revealed that the alloy powder contains ~0.25–0.3 mass% O, which indicates that a thin oxide layer is present on the investigated alloy powder particles. These oxide films could dissolve in the alloy melts during L-PBF. STEM characterization [21] has revealed that nanoscale oxide particles can be formed in the microstructure of as-built samples; however, the O content may be partially disbursed in the α-Al phase during solidification, resulting in the expanded lattice in the investigated L-PBF-built samples. A similar trend has been found in Al–2.5%Fe binary alloy samples [35]. However, direct evidence concerning the presence of O solute in the α-Al phase has not been obtained. Therefore, additional characterization experiments must be performed to clarify the state in which O exists in L-PBF-processed Al alloys.

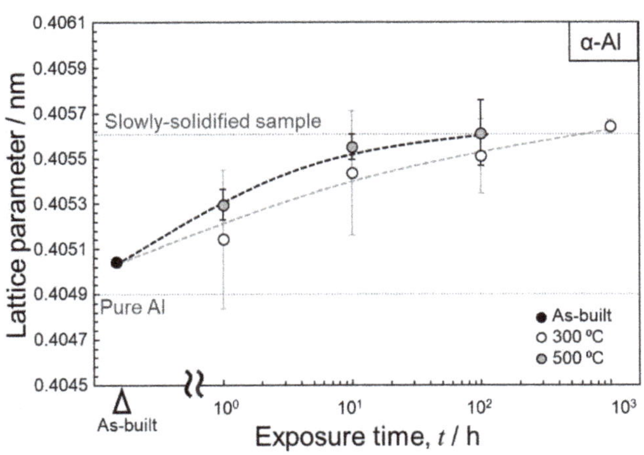

Figure 9. Changes in lattice parameter of the L-PBF-built Al–15%Fe alloy samples as a function of exposure time.

4. Conclusions

Variations in the refined Al–Fe intermetallic phases and microstructure of the α-Al matrix in an L-PBF-built Al–15%Fe alloy upon exposure to intermediate temperatures (300 and 500 °C) were examined. The key findings are summarized below.

- The microstructure of the as-built sample had a stable θ-$Al_{13}Fe_4$ phase localized along the melt-pool boundaries as well as numerous particles of the metastable Al_6Fe phase surrounded by a nanoscale α-Al/Al_6Fe cellular structure in the melt pools. The morphology of the θ phase remained almost unchanged even after 1000 h of exposure at 300 °C. The cellular-structured Al_6Fe phase dissolved in the α-Al matrix, which was followed by the growth of Al_6Fe-phase particles and the precipitation of the θ phase. The growth of the θ phase was more pronounced at the higher investigated temperature (500 °C).

- Numerous elongated grains with widths of a few tens of micrometers were observed in the α-Al microstructure of the as-built sample, in addition to locally present fine grains. Relatively equiaxed grains were often formed after exposure to 300 °C, particularly in the fine-grained regions. This tendency was more evident at the higher temperature of 500 °C, which led to the formation of a homogenous microstructure. The lattice parameter of the α-Al matrix increased with increasing exposure time, and almost stabilized at ~0.406 nm. The lattice expansion could be associated with the reduction of the Fe solute, which has a smaller atomic radius than that of Al.

Author Contributions: Conceptualization, N.T.; methodology, N.T. and W.W.; software, M.K. (Masaki Kato); validation, N.T., M.K. (Makoto Kobashi), and A.S.; formal analysis, W.W.; investigation, W.W. and N.T.; data curation, W.W. and N.T.; writing—original draft preparation, W.W.; writing—review and editing, N.T., A.S., and M.K. (Makoto Kobashi); visualization, W.W.; supervision, N.T.; project administration, N.T. and M.K. (Makoto Kobashi); funding acquisition, N.T. and M.K. (Makoto Kobashi). All authors have read and agreed to the published version of the manuscript.

Funding: This study was supported by the "Knowledge Hub Aichi," a Priority Research Project of the Aichi Prefectural Government, Japan; a JSPS KAKENHI grant (no. 20H02462; Japan); and the Light Metal Educational Foundation, Inc. (Japan).

Institutional Review Board Statement: Not applicable.

Informed Consent Statement: Not applicable.

Acknowledgments: The authors would like to thank Isao Murakami (Toyo Aluminum K. K.) for providing Al–15%Fe alloy powder.

Conflicts of Interest: The authors declare no conflict of interest.

References

1. Khorasani, A.; Gibson, I.; Veetil, J.K.; Ghasemi, A.H. A review of technological improvements in laser-based powder bed fusion of metal printers. *Int. J. Adv. Manuf. Technol.* **2020**, *108*, 191–209. [CrossRef]
2. Wei, C.; Liu, L.; Gu, Y.; Huang, Y.; Chen, Q.; Li, Z.; Li, L. Multi-material additive-manufacturing of tungsten-copper alloy bimetallic structure with a stainless-steel interlayer and associated bonding mechanisms. *Addit. Manuf.* **2021**, *50*, 102574. [CrossRef]
3. Fiocchi, J.; Tuissi, A.; Biffi, C. Heat treatment of aluminium alloys produced by laser powder bed fusion: A review. *Mater. Des.* **2021**, *204*, 109651. [CrossRef]
4. Zhou, L.; Hyer, H.; Chang, J.; Mehta, A.; Huynh, T.; Yang, Y.; Sohn, Y. Microstructure, mechanical performance, and corrosion behavior of additively manufactured aluminum alloy 5083 with 0.7 and 1.0 wt% Zr addition. *Mater. Sci. Eng. A* **2021**, *823*, 141679. [CrossRef]
5. Takata, N.; Kodaira, H.; Sekizawa, K.; Suzuki, A.; Kobashi, M. Change in microstructure of selectively laser melted AlSi10Mg alloy with heat treatments. *Mater. Sci. Eng. A* **2017**, *704*, 218–228. [CrossRef]
6. Liu, M.; Takata, N.; Suzuki, A.; Kobashi, M.; Kato, M. Enhancement in strength and ductility of laser powder bed fused Al–12Si alloy by introducing nanoscale precipitates. *Addit. Manuf. Lett.* **2021**, *1*, 100008. [CrossRef]
7. Suzuki, A.; Miyasaka, T.; Takata, N.; Kobashi, M.; Kato, M. Control of microstructural characteristics and mechanical properties of AlSi12 alloy by processing conditions of laser powder bed fusion. *Addit. Manuf.* **2021**, *48*, 102383. [CrossRef]
8. Wu, Y.; Zhang, T.; Chen, C.; Hosseini, S.R.E.; Zhang, X.; Zhou, K. Microstructure and mechanical property evolution of additive manufactured eutectic Al-2Fe alloy during solidification and aging. *J. Alloys Compd.* **2021**, *897*, 163243. [CrossRef]
9. Takata, N.; Liu, M.; Kodaira, H.; Suzuki, A.; Kobashi, M. Anomalous strengthening by supersaturated solid solutions of se-lectively laser melted Al–Si-based alloys. *Addit. Manuf.* **2020**, *33*, 101152.
10. Krakhmalev, P.; Vilardell, A.M.; Takata, N. Structural integrity I: Static mechanical properties. In *Fundamentals of Laser Powder Bed Fusion of Metals*, 1st ed.; Yadroitsev, I., Yadroitsava, I., Du Plessis, A., MacDonald, E., Eds.; Elsevier: Amsterdam, The Netherlands, 2021; pp. 349–376.
11. Qi, Y.; Hu, Z.; Zhang, H.; Nie, X.; Zhang, C.; Zhu, H. High strength Al–Li alloy development for laser powder bed fusion. *Addit. Manuf.* **2021**, *47*, 102249. [CrossRef]
12. Li, X.; Li, G.; Zhang, M.-X.; Zhu, Q. Novel approach to additively manufacture high-strength Al alloys by laser powder bed fusion through addition of hybrid grain refiners. *Addit. Manuf.* **2021**, *48*, 102400. [CrossRef]

13. Aboulkhair, N.T.; Simonelli, M.; Parry, L.; Ashcroft, I.; Tuck, C.; Hague, R. 3D printing of aluminium alloys: Additive manufacturing of aluminium alloys using selective laser melting. *Prog. Mater. Sci.* **2019**, *106*, 100578. [CrossRef]
14. Kotadia, H.; Gibbons, G.; Das, A.; Howes, P. A review of Laser Powder Bed Fusion Additive Manufacturing of aluminium alloys: Microstructure and properties. *Addit. Manuf.* **2021**, *46*, 102155. [CrossRef]
15. Yanase, Y.; Miyauchi, H.; Matsumoto, H.; Yokota, K. Densification Behavior and Microstructures of the Al–10%Si–0.35Mg Alloy Fabricated by Selective Laser Melting: From Experimental Observation to Machine Learning. *Mater. Trans.* **2022**, *63*, 176–184. [CrossRef]
16. Uzan, N.E.; Shneck, R.; Yeheskel, O.; Frage, N. High-temperature mechanical properties of AlSi10Mg specimens fabricated by additive manufacturing using selective laser melting technologies (AM-SLM). *Addit. Manuf.* **2018**, *24*, 257–263. [CrossRef]
17. Kimura, T.; Nakamoto, T.; Ozaki, T.; Miki, T. Microstructures and mechanical properties of aluminum-transition metal binary alloys (Al-Fe, Al-Mn, and Al-Cr) processed by laser powder bed fusion. *J. Alloys Compd.* **2021**, *872*, 159680. [CrossRef]
18. Michi, R.A.; Plotkowski, A.; Shyam, A.; Dehoff, R.R.; Babu, S.S. Towards high-temperature applications of aluminium alloys enabled by additive manufacturing. *Int. Mater. Rev.* **2021**, *67*, 298–345. [CrossRef]
19. Plotkowski, A.; Sisco, K.; Bahl, S.; Shyam, A.; Yang, Y.; Allard, L.; Nandwana, P.; Rossy, A.M.; Dehoff, R. Microstructure and properties of a high temperature Al–Ce–Mn alloy produced by additive manufacturing. *Acta Mater.* **2020**, *196*, 595–608. [CrossRef]
20. Wang, W.; Takata, N.; Suzuki, A.; Kobashi, M.; Kato, M. Processability and Optimization of Laser Parameters for Densification of Hypereutectic Al–Fe Binary Alloy Manufactured by Laser Powder Bed Fusion. *Crystals* **2021**, *11*, 320. [CrossRef]
21. Wang, W.; Takata, N.; Suzuki, A.; Kobashi, M.; Kato, M. Formation of multiple intermetallic phases in a hypereutectic Al–Fe binary alloy additively manufactured by laser powder bed fusion. *Intermetallics* **2020**, *125*, 106892. [CrossRef]
22. Walford, L.K. The structure of the intermetallic phase FeAl6. *Acta Cryst.* **1965**, *18*, 287–291. [CrossRef]
23. Wang, W.; Takata, N.; Suzuki, A.; Kobashi, M.; Kato, M. High-temperature strength sustained by nano-sized eutectic structure of Al–Fe alloy manufactured by laser powder bed fusion. *Mater. Sci. Eng. A* **2022**, *838*, 142782. [CrossRef]
24. de Araujo, A.P.M.; Pauly, S.; Batalha, R.L.; Coury, F.G.; Kiminami, C.S.; Uhlenwinkel, V.; Gargarella, P. Additive manufac-turing of a quasicrystal-forming Al95Fe2Cr2Ti1 alloy with remarkable high-temperature strength and ductility. *Addit. Manuf.* **2021**, *41*, 101960.
25. Pickens, J.R. High-strength aluminum powder metallurgy alloys. In *ASM Handbook, Volume 2: Properties and Selection: Non-Ferrous Alloys and Special-Purpose Materials*; ASM International: Ohio, OH, USA, 1990; pp. 200–215.
26. Bi, J.; Lei, Z.; Chen, Y.; Chen, X.; Tian, Z.; Lu, N.; Qin, X.; Liang, J. Microstructure, tensile properties and thermal stability of AlMgSiScZr alloy printed by laser powder bed fusion. *J. Mater. Sci. Technol.* **2020**, *69*, 200–211. [CrossRef]
27. Black, P.J. The structure of FeAl3. II. *Acta Crystallogr.* **1955**, *8*, 175–182. [CrossRef]
28. Takata, N.; Nishida, R.; Suzuki, A.; Kobashi, M.; Kato, M. Crystallographic features of microstructure in maraging steel fab-ricated by selective laser melting. *Metals* **2018**, *8*, 440. [CrossRef]
29. Aliravci, C.A.; Pekgüleryüz, M. Calculation of phase diagrams for the metastable Al-Fe phases forming in direct-chill (DC)-cast aluminum alloy ingots. *Calphad* **1998**, *22*, 147–155. [CrossRef]
30. Qi, X.; Takata, N.; Suzuki, A.; Kobashi, M.; Kato, M. Managing both high strength and thermal conductivity of a laser powder bed fused Al–2.5 Fe binary alloy: Effect of annealing on microstructure. *Mater. Sci. Eng. A* **2021**, *805*, 140591. [CrossRef]
31. Adam, C.; Hogan, L. Crystallography of the Al-Al3 Fe eutectic. *Acta Metall.* **1975**, *23*, 345–354. [CrossRef]
32. Chen, S.; Yang, D.; Yang, J.; Huang, J.; Zhao, X. Nanoscale structures of the interfacial reaction layers between molten alu-minium and solid steel based on thermophysical simulations. *J. Alloys Compd.* **2018**, *739*, 184–189. [CrossRef]
33. Teatum, E.T.; Gschneidner, K.A., Jr.; Waber, J.T. *Compilation of Calculated Data Useful in Predicting Metallurgical Behavior of the Elements in Binary Alloy Systems*; Los Alamos Scientific Laboratory of the University of California: Los Alamos, NM, USA, 1968. [CrossRef]
34. Takata, N.; Okano, T.; Suzuki, A.; Kobashi, M. Microstructure of intermetallic-reinforced Al-Based alloy composites fabricated using eutectic reactions in Al–Mg–Zn ternary system. *Intermetallics* **2018**, *95*, 48–58. [CrossRef]
35. Qi, X.; Takata, N.; Suzuki, A.; Kobashi, M.; Kato, M. Change in microstructural characteristics of laser powder bed fused Al–Fe binary alloy at elevated temperature. *J. Mater. Sci. Technol.* **2021**, *97*, 38–53. [CrossRef]

materials

MDPI

Article

Correlation between Differential Fast Scanning Calorimetry and Additive Manufacturing Results of Aluminium Alloys

Olaf Kessler [1,2,*], Evgeny Zhuravlev [1,2], Sigurd Wenner [3], Steffen Heiland [4] and Mirko Schaper [4]

1 Chair of Materials Science, Rostock University, 18051 Rostock, Germany
2 Competence Centre °CALOR, Department Life, Light and Matter, Rostock University, 18051 Rostock, Germany
3 SINTEF Stiftelsen for Industriell og Teknisk Forskning, 7465 Trondheim, Norway
4 Chair of Materials Science, Paderborn University, 33098 Paderborn, Germany
* Correspondence: olaf.kessler@uni-rostock.de

Abstract: High-strength aluminium alloy powders modified with different nanoparticles by ball milling (7075/TiC, 2024/CaB6, 6061/YSZ) have been investigated in-situ during rapid solidification by differential fast scanning calorimetry (DFSC). Solidification undercooling has been evaluated and was found to decrease with an increasing number of nanoparticles, as the particles act as nuclei for solidification. Lower solidification undercooling of individual powder particles correlates with less hot cracking and smaller grains in the material produced by powder bed fusion of metals by a laser beam (PBF-LB/M). Quantitatively, solidification undercooling less than about 10–15 K correlates with almost crack-free PBF-LB/M components and grain sizes less than about 3 μm. This correlation shall be used for future purposeful powder material design on small quantities before performing extensive PBF-LB/M studies.

Keywords: PBF-LB/M; aluminium alloys; hot cracking; rapid solidification; differential fast scanning calorimetry; undercooling; grain size; crack density

Citation: Kessler, O.; Zhuravlev, E.; Wenner, S.; Heiland, S.; Schaper, M. Correlation between Differential Fast Scanning Calorimetry and Additive Manufacturing Results of Aluminium Alloys. *Materials* **2022**, *15*, 7195. https://doi.org/10.3390/ma15207195

Academic Editor: Amir Mostafaei

Received: 5 September 2022
Accepted: 10 October 2022
Published: 15 October 2022

1. Introduction

Powder bed fusion of metals by a laser beam (PBF-LB/M) is a very attractive production process due to its high flexibility and a high degree of geometrical freedom. Local microstructures and properties of PBF-LB/M components are strongly influenced by process inherent complex temperature/time profiles with multiple rapid melting/solidification cycles and multiple rapid heating/cooling cycles. Considering lightweight applications, PBF-LB/M is very promising for high-strength aluminium alloys. Unfortunately, PBF-LB/M of high-strength aluminium alloys is strongly affected by hot cracking. When cooling along the relatively large solidification intervals (e.g., about 80 K in equilibrium for alloy 7021 [1]), the remaining melt is encapsulated between the growing dendrites. During the final solidification, these melt areas shrink. If this shrinkage cannot be accommodated by deformation of the surrounding solid network, hot cracks can occur [2–4]. One successful approach to overcome hot cracking during PBF-LB/M of high-strength aluminium alloys is the addition of high melting point nanoparticles to the aluminium alloy powders. Aluminium alloy powder particles for PBF-LB/M are typically in the 10–60 μm range. Throughout the whole paper, we will consequently differentiate between the terms "nanoparticles" for the additives and "particles" for the PBF-LB/M powders. Ball milling is one suitable method to add the nanoparticles to the aluminium powder. These nanoparticles act as solidification nuclei and suppress hot cracking. Table 1 gives an overview about several successful investigations regarding utilised aluminium alloys and nanoparticles. In some cases, the added nanoparticles act as nuclei themselves, and in other cases (Zr, Ti), the nanoparticles first react with the aluminium melt to form intermetallic compounds (Al_3Zr, Al_3Ti), which then act as nuclei.

In all these investigations, the nanoparticle types have been carefully chosen by suitable low crystal lattice mismatch between nuclei and aluminium matrix. However, the nanoparticles sizes, amounts and addition methods have been selected based on extensive experimental PBF-LB/M studies, requiring large amounts of modified powders (typically several kilograms). Our objective is to design promising powder modifications in small quantities (a few grams) before performing extensive PBF-LB/M studies. Therefore, we analyse the rapid melting/solidification behaviour of the modified powders in-situ by calorimetry. Especially differential fast scanning calorimetry (DFSC) with sample sizes of a few 10 µm particles (equivalent to PBF-LB/M powder particle sizes) and heating/cooling rates of up to 10^6 K/s (equivalent to PBF-LB/M processes) is a suitable method [5]. In detail, rapid solidification conditions of single powder particles by DFSC differ from those in PBF-LB/M melt pools with dimensions in the several 100 µm ranges, where numerous powder particles are molten simultaneously. Nevertheless, we presume, that the main solidification behaviour can be approximated. Rapid solidification characteristics, e.g., undercooling, will then be correlated with grain sizes and crack characteristics in belonging PBF-LB/M components. This correlation shall be used for future purposeful powder material designs tailored for PBF-LB/M.

Table 1. Overview of several successful investigations regarding used aluminium alloys and nanoparticles.

Reference	Al Alloy	Nanoparticles
Gu et al., 2014 [6]	AlSi10Mg	TiC
Martin et al., 2017 [7]	7075	$Zr \rightarrow Al_3Zr$
Tan et al., 2020 [8]	2024	$Ti \rightarrow Al_3Ti$
Zhao et al., 2020 [9]	5024	TiC
Xi et al., 2020 [10]	AlSi10Mg	TiB_2
Opprecht et al., 2020 [11]	6061	YSZ * (60 nm) $\rightarrow Al_3Zr$
Zhuravlev et al., 2021 [5]	7075	TiC (40 nm)
Heiland et al., 2021 [12]	7075	TiC (40 nm)
Mair et al., 2022 [13]	2024	CaB_6 (200 nm)

* Yttrium stabilized zirconia $Zr_{(1-x)}Y_xO_2$.

2. Materials and Methods

Table 2 contains the investigated aluminium alloys, nanoparticles and references. Nanoparticle amounts have been varied between zero and a few mass %. The rapid melting/solidification behaviour of several aluminium powders modified by different nanoparticles has been analysed in-situ by DFSC. Figure 1 shows a schematic DFSC sensor (side view) as well as a light microscope image (top view). The samples are individual powder particles with diameters of about 20 µm. Heating and cooling rates of 10^4 K/s and maximum temperatures up to 823 °C have been used. DFSC measurements require careful temperature correction related to suitable reference temperatures. Details on measurement and evaluation have been published in [5]. Figure 2b shows three typical DFSC heating and cooling curves on alloy 2024 at a rate of 10^4 K/s without and with CaB_6 nanoparticles. The three bottom curves (endothermal) belong to melting. During rapid heating, we can see incipient melting slightly above 500 °C and a continuous ongoing melting up to melting finish at about 700 °C almost identical for all three variants. Melting finish depending on the heating rate has been extrapolated to heating rate zero, i.e., liquidus temperature [5] and is taken as reference temperature for undercooling ($T_{m,0}$ = 638 °C dashed line). The three upper curves (exothermal) belong to solidification. Solidification of 2024 starts about 570 °C, i.e., undercooling amounts about 70 K, whereas solidification of 2024/0.3% CaB_6 starts about 610 °C, i.e., undercooling amounts only about 30 K. In this example, the decrease of undercooling due to the nanoparticles is about 40 K. For the variant 2024/0.5% CaB_6 rapid solidification starts almost without any undercooling. Figure 2a,c shows very similar behaviour in DFSC of single particles from alloys 7075/TiC and 6061/YSZ. Nanoparticle addition decreases solidification undercooling.

Table 2. Investigated aluminium alloys, nanoparticles and references.

Reference	Al-Alloy	Nanoparticles, Amount	NP Deposition	DFSC Rate
Zhuravlev et al., 2021 [5]	7075	TiC 40nm 0–0.5 mass %	ball milling wet deposition	10^4 K/s
Heiland et al., 2021 [12]	7075	TiC 40 nm 0–2.5 mass %	ball milling	10^4 K/s
this work	7021	TiC 40 nm 0–1.75 mass %	ball milling	10^4 K/s
this work	7021	TiB_2 50 nm 0–1.75 mass %	ball milling	10^4 K/s
Mair et al., 2022 [13]	2024	CaB_6 200 nm 0–2 mass %	ball milling	10^4 K/s
Opprecht et al., 2020 [11]	6061	YSZ 60 nm 0–4 volume % *	ball milling	10^3 K/s

* Density of YSZ is about 6 g/cm^3, i.e., roughly double the density of aluminium.

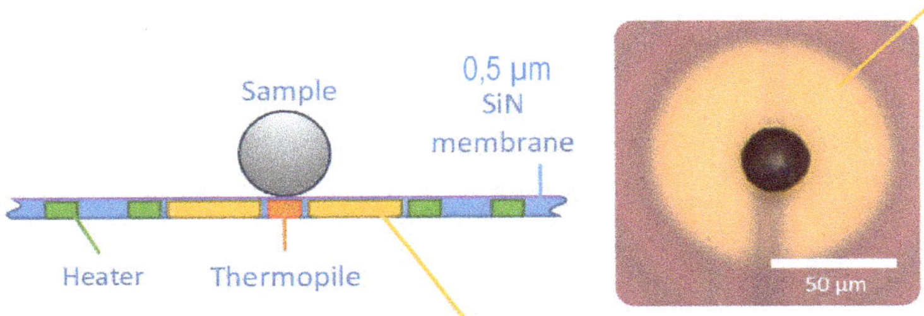

Figure 1. Schematic of the DFSC sensor (side view) as well as a light micrograph (top view) [5].

Whereas the rapid melting curves of 2024/CaB6 and 6061/YSZ (Figure 2b,c) are very similar, they differ for 7075/TiC (Figure 2a). This effect can arise from slightly different powder particle masses, from slightly different powder particle positions on the sensor and from different thermal contacts between powder particles and sensors. All DFSC curves have been temperature corrected accordingly [5], i.e., rapid solidification onset can be determined properly.

Each powder variant without and with nanoparticles has been analysed by DFSC on several individual powder particles. On each individual powder particle, several repeated heating/cooling cycles have been performed. Previous work has shown, that up to 300 repeated heating/cooling cycles can be performed on one individual aluminium powder particle without changing its melting/solidification behaviour [14]. Table 3 contains the number of analysed particles, number of repetitions per particle and the numbers of evaluated DFSC experiments for each variant. Typical undercooling scatter ranges from about ±20 K for high undercooling without nanoparticles to about ±2 K for low undercooling with nanoparticles.

Table 3. Numbers of analysed particles, numbers of repetitions per particle and numbers of evaluated DFSC experiments for each variant.

Alloy/Nanoparticles	Numbers of Analysed Particles	Numbers of Repetitions per Particle	Numbers of Evaluated DFSC Experiments
7075/TiC	at least 5	about 60	about 300
2024/CaB_6	at least 3	about 75	about 225
6061/YSZ	at least 15	about 10	about 150

Exactly the same nanoparticle-modified powder batches as in DFSC were used in PBF-LB/M processes. PBF-LB/M results and parameters are given in the references [5,11–13]. Additionally, the two powder variants 7021/TiC and 7021/TiB_2 have been investigated with both methods in this work. Here, PBF-LB/M took place on an SLM 250^{HL} machine (SLM Solutions Group AG, Lübeck, Germany), equipped with a YLM-400-WC Laser (IPG

Photonics, Oxford, MA, USA) by the following parameters: layer thickness 0.05 mm, hatch distance 0.08 mm, scan rate 900 mm/s, laser power 370 W, resulting in volume energy of 102.8 W/mm^3. The specimens were fabricated under an argon atmosphere with a residual oxygen level of approximately 2000 ppm. Own data from [5,12] has also been re-evaluated regarding the above parameter set, i.e., all results on 7075 and 7021 shown below originate from identical PBF-LB/M parameters.

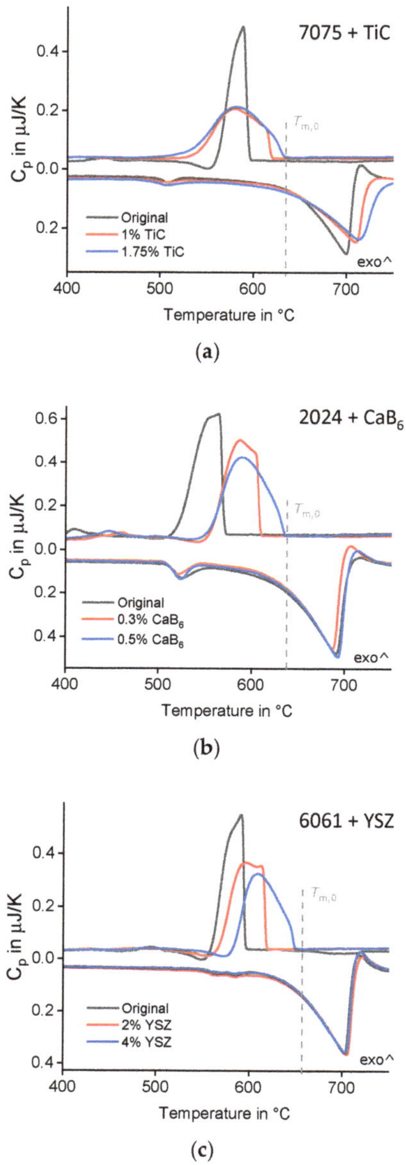

Figure 2. Exemplary DFSC heating and cooling curves (**a**) alloy 7075 without and with TiC nanoparticles, rate 10^4 K/s. (**b**) Alloy 2024 without and with CaB_6 nanoparticles, rate 10^4 K/s. (**c**) alloy 6061 without and with YSZ nanoparticles, rate 10^3 K/s. The vertical axis shows heat capacity in µJ/K, exothermal reactions upwards.

As-build samples have been analysed regarding grain size and crack characteristics by metallographic methods. The metallographic methods are also described in the references [5,11–13]. In these experiments, cracks have been described by different measures, i.e., crack density determined on cross sections by light microscopy [5,12], crack volume determined by X-ray microtomography [13] and total crack length per area determined on cross sections by light microscopy [11]. To compare these different crack measures, we suggest a crack characteristic value C, which is defined by the ratio of the crack measure with nanoparticles to the maximum crack measure without nanoparticles, equations (1–3). In further evaluation, we propose that these crack characteristic values C_D, C_V and C_L can be directly compared, i.e., we call them just crack characteristic value C. By this definition the crack characteristic value C can exist in the range of 0 to 1, with C = 0 meaning complete crack suppression by nanoparticles and C = 1 meaning no change in cracking with nanoparticles. This relative crack characteristic C can be used to compare different crack measures, different aluminium alloys, different nanoparticles and different PBF-LB/M processes.

$$C_D = \frac{(crack\ density)_{with\ NP}}{(crack\ density)_{without\ NP}} \tag{1}$$

for 7075/TiC, 7021/TiC, 7021/TiB_2 data from [5,12] and this work

$$C_V = \frac{(crack\ volume)_{with\ NP}}{(crack\ volume)_{without\ NP}} \tag{2}$$

for 2024/CaB_6, data from [13].

$$C_L = \frac{(total\ crack\ length)_{with\ NP}}{(total\ crack\ length)_{without\ NP}} \tag{3}$$

for 6061/YSZ, data from [11].

Single rapidly solidified particles from DFSC were analysed by SEM and TEM. The metallographic preparation route for SEM analysis of such small and individual powder particles has been developed as described in [15]. For TEM investigations of solidified particles, the particles were first embedded in epoxy, then a FEI Helios G4 focused ion beam (FIB) instrument was used to cut a thin section through the particle using the standard lift-out approach. A double aberration corrected JEOL ARM-200F was used for scanning transmission electron microscopy (STEM) and elemental mapping of the particle surface with electron energy loss spectroscopy (EELS).

3. Results and Discussion

DFSC solidification undercooling and PBF-LB/M crack characteristics C, as well as grain size have been investigated for each individual variant of aluminium alloy, nanoparticle type and nanoparticle amount according to Table 2. All powder variants will be compared in terms of crack characteristics C, grain size and solidification undercooling.

3.1. Correlation between Crack Characteristics C and Solidification Undercooling

Figure 3 shows the correlation between crack characteristics C and solidification undercooling. Each individual point displays crack characteristics and solidification undercooling for one powder variant. Results from the same aluminium alloy/nanoparticle system with different nanoparticle amounts are plotted in the same colour. Data points are labelled with the belonging nanoparticle amounts in mass %. In most cases, undercooling, as well as crack characteristics decrease with increasing nanoparticle amount. The only exception is the powder variant 7021/TiB_2, which even at a high amount of 1.75 mass % reduced undercooling only to 16 K and crack characteristic value only to about 0.4.

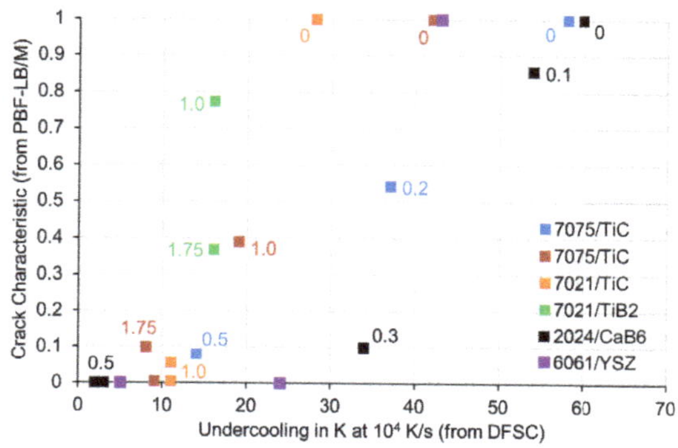

Figure 3. Correlation between crack characteristics C and solidification undercooling. Data points are labelled with the belonging nanoparticle amounts in mass %. Relative crack characteristics of 2024/CaB_6 and 6061/YSZ have been calculated from [11,13].

In Figure 4, we have added arrows as a guide for the eye, each arrow belonging to one aluminium alloy/nanoparticle system. Now it can be clearly seen, that crack characteristic C decreases with decreasing undercooling for all variants. For a given aluminium alloy/nanoparticle system, undercooling decreases with an increasing number of nanoparticles. The effect of undercooling on crack characteristics depends on the individual aluminium alloy/nanoparticle system, which can be explained by the different PBF-LB/M machines and parameters as well as crack measures used. In the bottom left corner of the diagram, we find a successful process window relative consistent for all investigated variants. A low solidification undercooling of less than about 10–15 K in DFSC (at a cooling rate of 10^4 K/s) correlates with almost crack-free PBF-LB/M components. This correlation can be used in future for the purposeful design of powder materials in small quantities (a few grams) before conducting extensive PBF-LB/M studies.

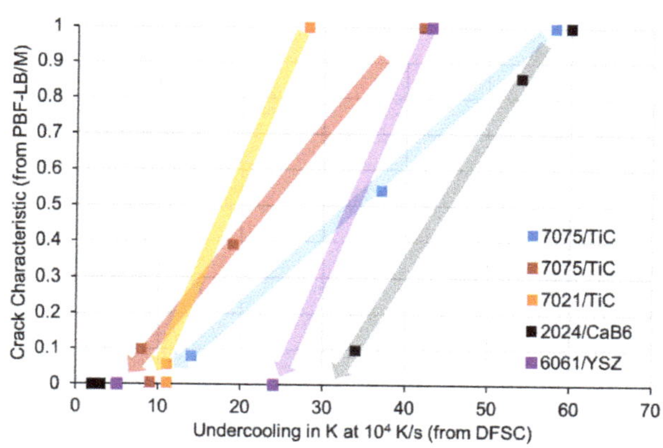

Figure 4. Correlation between crack characteristics C and solidification undercooling with arrows as guides for the eye. Relative crack characteristics of 2024/CaB_6 and 6061/YSZ have been calculated from [11,13].

3.2. Correlation between Grain Size and Solidification Undercooling

Figure 5 shows the correlation between grain size and solidification undercooling. Each individual point displays grain size and solidification undercooling for one powder variant. Please have in mind, that grain sizes result from PBF-LB/M samples and therefore can be larger than the individual powder particles in DFSC. Results from the same aluminium alloy/nanoparticle system with different nanoparticle amounts are plotted in the same colour. Like crack characteristic C, grain size decreases with decreasing undercooling, proving that the nanoparticles are responsible for efficient inoculation. Besides decreasing grain size, also grain geometry changes from columnar to equiaxed with the increasing number of nanoparticles [5,11–13]. In the case of columnar grains, the column widths have been plotted in Figure 5. Column lengths have grown even larger in the range of several 100 µm. In the bottom left corner of the diagram, we find again a successful process window relative consistent for all investigated variants. A low solidification undercooling of less than about 10–15 K in DFSC (at a cooling rate of 10^4 K/s) correlates with low grain sizes of less than about 3 µm.

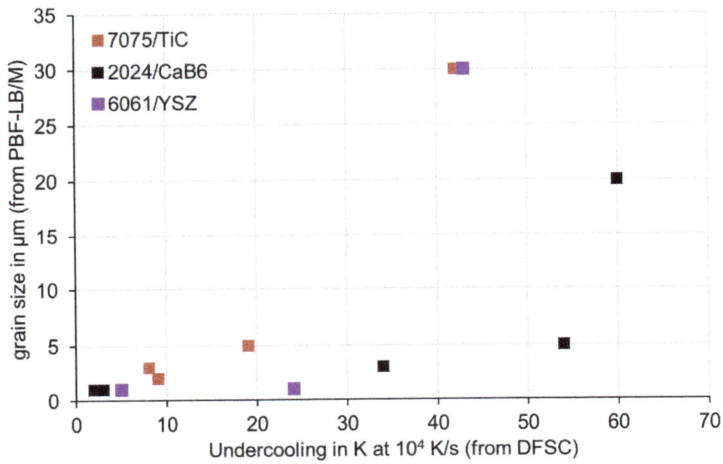

Figure 5. Correlation between grain size and solidification undercooling. Grain sizes of 2024/CaB_6 and 6061/YSZ have been adopted from [11,13].

3.3. *Correlation between Crack Characteristics C and Grain Size*

Finally, we have correlated crack characteristics C and grain size independent of the alloy/nanoparticle system (Figure 6). Some data points crack characteristic/grain size overlap in this diagram and have been marked accordingly (2×, 4×). As expected, crack characteristic C decreases with decreasing grain size. Moreover, in this diagram we find a successful process window in the bottom left corner which is relatively consistent for all investigated variants. Low grain size of less than about 3 µm correlates with almost crack-free PBF-LB/M components. As mentioned above, grain shape changes concurrently with grain size from large columnar grains to small equiaxed grains.

This correlation is in good agreement with the accepted mechanism of hot cracking during solidification [2–4]. Large columnar dendrites form a solid network early during solidification, which encloses the remaining encapsulated melt volumes. During further solidification of these encapsulated melt volumes, their shrinkage cannot be sufficiently accommodated by the deformation of the solid network and hot cracking occurs in these places. Small equiaxed dendrites form such a solid network significantly later during solidification. The remaining melt volumes are not encapsulated early. Instead, they are interconnected and can balance volume shrinkage. This correlation confirms our presented approach, to use DFSC on small quantities (a few grams) for future purposeful design of powder materials for crack-free PBF-LB/M components.

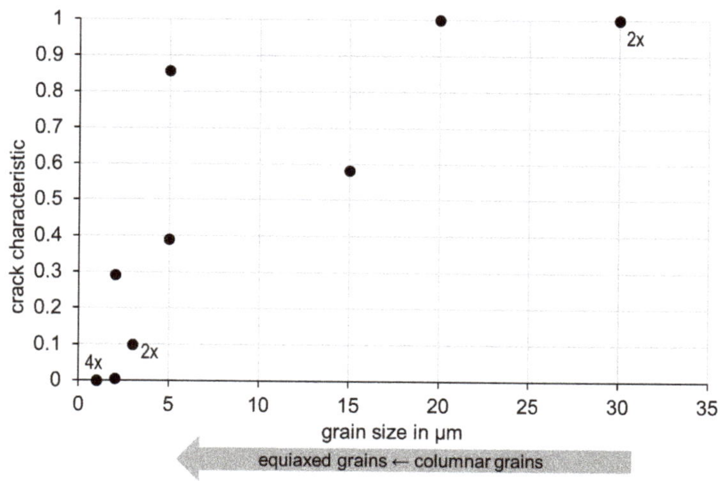

Figure 6. Correlation between crack characteristics C and grain size. Arrow indicates concurrent change from large columnar to small equiaxed grains. Some data points overlap in this diagram and have been marked accordingly (2×, 4×).

3.4. *Efficiency of Nanoparticle Inoculation*

Further, we make an estimation of the inoculation efficiency of nanoparticles on powder particle surfaces during PBF-LB/M. Table 4 gives the necessary amounts of nanoparticles for successful PBF-LB/M to suppress hot cracks in the present work, as well as from [10,12]. In many cases, about a few mass % of nanoparticles are required. The following calculation illustrates the consequences. Let us assume a typical aluminium particle diameter for PBF-LB/M of d_p = 40 μm and a typical nanoparticle dimension of d_{np} = 40 nm (Table 2). Let us further assume a continuous monolayer of nanoparticles on the aluminium powder particle surface (Figure 7 left). In this case, the content of nanoparticles c_{np} in mass % can be approximated by Equation (4) considering the particle density ρ_p = 2.7 g/cm^3 (aluminium) and the nanoparticle density ρ_{np} = 4.9 g/cm^3 (TiC).

$$c_{np} = \frac{\rho_{np}}{\rho_p} \cdot \frac{\left(d_p + 2 \cdot d_{np}\right)^3 - d_p^3}{d_p^3} \quad (4)$$

Table 4. The necessary content of nanoparticles for successful PBF-LB/M to suppress hot cracks.

Reference	Alloy/Nanoparticles	Necessary Amount of Nanoparticles
this work	7075/TiC	1.75 mass %
this work	7021/TiC	1 mass %
Mair et al., 2022 [13]	2024/CaB_6	0.5 mass %
Opprecht et al., 2020 [11]	6061/YSZ	2 volume % *

* Density of YSZ is about 6 g/cm^3, i.e., roughly double the density of aluminium.

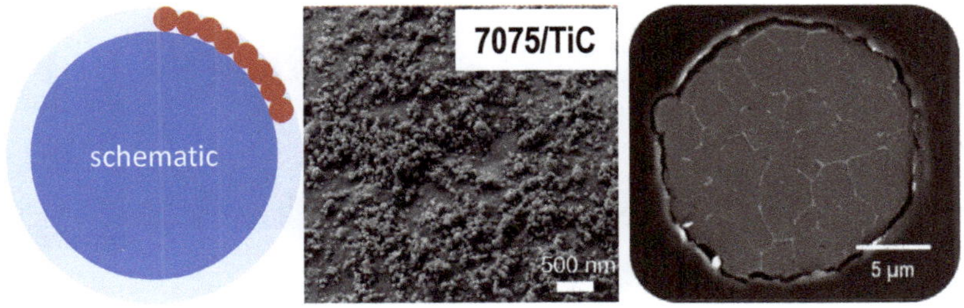

Figure 7. (**Left**) Schematic of a monolayer of nanoparticles. (**Middle**) Surface of a powder particle 7075/TiC. (**Right**) Cross section of an individual powder particle 7075/TiC, rapidly solidified by DFSC.

Under the mentioned assumptions, one monolayer of TiC nanoparticles on aluminium alloy 7075 particles equals an amount of about 0.6 volume%, corresponding to about 1.0 mass % nanoparticles. Figure 7 (middle) shows an SEM image of the surface of a 7075/TiC powder particle with 1 mass % TiC. The powder particle surface is almost completely covered by nanoparticles, which proves that the above approximation is realistic. We can see some gaps, but in other areas, the nanoparticles build multilayers. Next, we estimate the number of nanoparticles N_{np} in the monolayer according to Equation (5).

$$N_{np} = \frac{\left(d_p + 2 \cdot d_{np}\right)^3 - d_p^3}{d_{np}^3} \tag{5}$$

This estimation results in about 10^6 nanoparticles per aluminium powder particle, i.e., about 10^6 potential nuclei for solidification. For comparison, Figure 7 (right) shows a SEM image of a metallographic cross section of one individual powder particle of 7075/TiC with a diameter of about 20 µm, solidified by DFSC with a cooling rate of 10^4 K/s. The dendrite arm size respectively grain size amounts to a few µm and corresponds very well to the grain size in crack-free PBF-LB/M components (Figure 6). This is another indication, that rapid single-particle solidification by DFSC resembles melt pool conditions in PBF-LB/M. In Figure 7 right, we can only very roughly guess the number of efficient solidification nuclei, especially because we see a certain number of dendrite arms, but we do not know their crystal orientation. However, even if we very generously guess the number of active nuclei to be in the order of 10 to 100, we can conclude on a very low nucleation efficiency of about 10^{-5} to 10^{-4} for such nanoparticles added by ball milling.

The reason for this low nucleation efficiency is unclear at the moment. One assumption is, that the nanoparticles added by ball milling stick to the relative stable natural oxide layer of the aluminium powder particles, which may hinder direct contact with the aluminium melt. This assumption is supported by the STEM/EELS cross-section image in Figure 8, which has been prepared from a single 7075/TiC particle after rapid solidification by DFSC. We can clearly see the TiC nanoparticles with dimensions of about 40 nm as well as the oxide layer in between. This assumption needs further investigation, but in case it is realistic, it would require other, more effective inoculation methods, i.e., adding nanoparticles inside the powder particle volumes [16]. This could drastically reduce the number of nanoparticles required to achieve a reduced solidification undercooling.

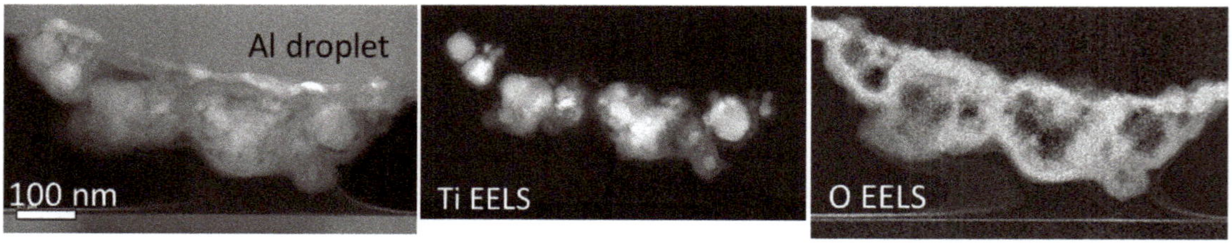

Figure 8. STEM/EELS cross section image, which has been prepared by FIB from a single 7075/TiC particle after rapid solidification by DFSC.

4. Conclusions

Different high-strength aluminium alloy powders modified with different nanoparticles by ball milling (7075/TiC, 2024/CaB_6, 6061/YSZ) have been investigated in-situ during rapid solidification by DFSC. Solidification undercooling has been evaluated and was found to decrease with increasing number of nanoparticles. Solidification undercooling from DFSC was compared with PBF-LB/M results from the same powder batches regarding hot cracking and grain size. Low solidification undercooling correlates with little hot cracking and small grains. Quantitatively, solidification undercooling less than about 10–15 K correlates with almost crack free PBF-LB/M components and grain sizes less than about 3 μm. This correlation will be used for future purposeful powder material design on small quantities before performing extensive PBF-LB/M studies.

Nanoparticles added on powder particle surfaces by ball milling exhibit a very low nucleation efficiency during rapid solidification in the range of 10^{-5} to 10^{-4} (active nuclei related to total number of nanoparticles). The reason for this low nucleation efficiency shall be further investigated and other inoculation methods instead of ball milling shall be considered for higher nucleation efficiency such that the amount of added nanoparticles can be greatly reduced while retaining the improved properties of PBF-LB/M components.

Author Contributions: O.K.: conceptualization, methodology, evaluation, writing, supervision; E.Z.: calorimetry experiments, S.W.: transmission electron microscopy, S.H.: additive manufacturing experiments, M.S.: conceptualization, supervision. All authors have read and agreed to the published version of the manuscript.

Funding: The authors gratefully thank the German Research Foundation (DFG) for funding this investigation within the special priority program (SPP 2122), "Materials for Additive Manufacturing", Grant-ID KE616/27-1 and -2, SCHA 1484/43-1 and -2, ZH661/3-1 and -2.

Institutional Review Board Statement: Not applicable.

Informed Consent Statement: Not applicable.

Acknowledgments: We gratefully acknowledge the modified powder supply and fruitful discussions: Powder 2024/CaB_6 by Philipp Mair and Gerhard Leichtfried from Faculty of Engineering Sciences, University of Innsbruck, Department of Mechatronics, Materials Science, Innsbruck, Austria. Powder 6061/YSZ by Mathieu Opprecht and Guilhem Roux from Univ Grenoble Alpes, CEA, LITEN, DTNM, Grenoble, France. The FIB work was done within the NorFab facility (Research Council of Norway (RCN) grant 245963/F50), and the TEM work was done using the NORTEM infrastructure (RCN grant 197405), both at the Norwegian University of Science and Technology (NTNU).

Conflicts of Interest: The authors declare no conflict of interest.

References

1. Dimitrii, A. *Petrov and MSIT®. SpringerMaterials*; Effenberg, G., Ed.; 10.11491.1.9; Springer-Verlag GmbH: Heidelberg, Germany, 1993.
2. Kurz, W.; Fisher, D.J. *Fundamentals of Solidification*; Trans Tech Publications Ltd.: Wollerau, Switzerland, 1998.
3. Coniglio, N.; Cross, C.E. Initiation and growth mechanisms for weld solidification cracking. *Int. Mater. Rev.* **2013**, *58*, 375–397. [CrossRef]
4. Rappaz, M.; Drezet, J.; Gremaud, M. A new hot-tearing criterion. *Metall. Mater. Trans. A* **1999**, *30*, 449–455. [CrossRef]

5. Zhuravlev, E.; Milkereit, B.; Yang, B.; Heiland, S.; Vieth, P.; Voigt, M.; Schaper, M.; Grundmeier, G.; Schick, C.; Kessler, O. Assessment of AlZnMgCu alloy powder modification for crack-free laser powder bed fusion by differential fast scanning calorimetry. *Mater. Des.* **2021**, *204*, 109677. [CrossRef]
6. Gu, D.; Wang, H.; Chang, F.; Dai, D.; Yuan, P.; Hagedorn, Y.-C.; Meiners, W. Selective laser melting additive manufacturing of TiC/AlSi10Mg bulk-form nanocomposites with tailored microstructures and properties. *Phys. Procedia* **2014**, *56*, 108–116. [CrossRef]
7. Martin, J.H.; Yahata, B.D.; Hundley, J.M.; Mayer, J.A.; Schaedler, T.A.; Pollock, T.M. 3D printing of high-strength aluminium alloys. *Nature* **2017**, *549*, 365–369. [CrossRef] [PubMed]
8. Tan, Q.; Zhang, J.; Sun, Q.; Fan, Z.; Li, G.; Yin, Y.; Liu, Y.; Zhang, M.-X. Inoculation treatment of an additively manufactured 2024 aluminium alloy with titanium nanoparticles. *Acta Mater.* **2020**, *196*, 1–16. [CrossRef]
9. Zhao, T.; Dahmen, M.; Cai, W.; Alkhayat, M.; Schaible, J.; Albus, P.; Zhong, C.; Hong, C.; Biermann, T.; Zhang, H.; et al. Laser metal deposition for additive manufacturing of AA5024 and nanoparticulate TiC modified AA5024 alloy composites prepared with balling milling process. *Opt. Laser Technol.* **2020**, *131*, 106438. [CrossRef]
10. Xi, L.; Gu, D.; Guo, S.; Wang, R.; Ding, K.; Prashanth, K.G. Grain refinement in laser manufactured Al-based composites with TiB2 ceramic. *J. Mater. Res. Technol.* **2020**, *9*, 2611–2622. [CrossRef]
11. Opprecht, M.; Garandet, J.-P.; Roux, G.; Flament, C.; Soulier, M. A solution to the hot cracking problem for aluminium alloys manufactured by laser beam melting. *Acta Mater.* **2020**, *197*, 40–53. [CrossRef]
12. Heiland, S.; Milkereit, B.; Hoyer, K.-P.; Zhuravlev, E.; Kessler, O.; Schaper, M. Requirements for successfully processing high-strength AlZnMgCu alloys with PBF-LB/M. *Materials* **2021**, *14*, 7190. [CrossRef] [PubMed]
13. Mair, P.; Kaserer, L.; Braun, J.; Stajkovic, J.; Klein, C.; Schimbäck, D.; Perfler, L.; Zhuravlev, E.; Kessler, O.; Leichtfried, G. Dependence of mechanical properties and microstructure on solidification onset temperature for varying Al2024/CaB_6 mixtures processed using laser powder bed fusion. *Mater. Sci. Eng. A* **2022**, *833*, 142552. [CrossRef]
14. Peng, Q.; Yang, B.; Milkereit, B.; Liu, D.; Springer, A.; Rettenmayr, M.; Schick, C.; Keßler, O. Nucleation Behavior of a Single Al-20Si Particle Rapidly Solidified in a Fast Scanning Calorimeter. *Materials* **2021**, *14*, 2920. [CrossRef] [PubMed]
15. Milkereit, B.; Meißner, Y.; Ladewig, C.; Osten, J.; Peng, Q.; Yang, B.; Springer, A.; Keßler, O. Metallographic Preparation of Single Powder Particles. *Pract. Metallogr.* **2021**, *58*, 129–139. [CrossRef]
16. Hengsbach, F.; Hoyer, K.-P.; Schaper, M.; Andreiev, A. Isotropic, Crack-Free Steel Design Using an Additive Manufacturing Method. Patent WO 2021/180777 A1, 16 September 2021.

materials

MDPI

Article

Quality over Quantity: How Different Dispersion Qualities of Minute Amounts of Nano-Additives Affect Material Properties in Powder Bed Fusion of Polyamide 12

Alexander Sommereyns [1,2,*], Stan Gann [3], Jochen Schmidt [4,5], Abootorab Baqerzadeh Chehreh [6], Arne Lüddecke [7], Frank Walther [6], Bilal Gökce [3,8], Stephan Barcikowski [3] and Michael Schmidt [1,2]

1 Institute of Photonic Technologies (LPT), Friedrich-Alexander-Universität Erlangen-Nürnberg, Konrad-Zuse-Str. 3/5, 91052 Erlangen, Germany; michael.schmidt@lpt.uni-erlangen.de
2 Erlangen Graduate School in Advanced Optical Technologies (SAOT), Friedrich-Alexander-Universität Erlangen-Nürnberg, Paul-Gordan-Str. 6, 91052 Erlangen, Germany
3 Technical Chemistry I and Center for Nanointegration Duisburg-Essen (CENIDE), University of Duisburg-Essen, Universitaetsstr. 7, 45141 Essen, Germany; stan.gann@uni-due.de (S.G.); goekce@uni-wuppertal.de (B.G.); stephan.barcikowski@uni-due.de (S.B.)
4 Institute of Particle Technology, Friedrich-Alexander-Universität Erlangen-Nürnberg, Cauerstr. 4, 91058 Erlangen, Germany; jochen.schmidt@fau.de
5 Interdisciplinary Center for Functional Particle Systems, Friedrich-Alexander-Universität Erlangen-Nürnberg, Haberstr. 9a, 91058 Erlangen, Germany
6 Department of Materials Test Engineering (WPT), TU Dortmund University, Baroper Str. 303, 44227 Dortmund, Germany; abootorab.chehreh@tu-dortmund.de (A.B.C.); frank.walther@tu-dortmund.de (F.W.)
7 Institute for Particle Technology, Technische Universität Braunschweig, Volkmaroder Str. 5, 38104 Braunschweig, Germany; a.lueddecke@tu-bs.de
8 Materials Science and Additive Manufacturing, School of Mechanical Engineering and Safety Engineering, University of Wuppertal, Gaußstr. 20, 42119 Wuppertal, Germany
* Correspondence: alexander.sommereyns@lpt.uni-erlangen.de

Citation: Sommereyns, A.; Gann, S.; Schmidt, J.; Chehreh, A.B.; Lüddecke, A.; Walther, F.; Gökce, B.; Barcikowski, S.; Schmidt, M. Quality over Quantity: How Different Dispersion Qualities of Minute Amounts of Nano-Additives Affect Material Properties in Powder Bed Fusion of Polyamide 12. *Materials* **2021**, *14*, 5322. https://doi.org/10.3390/ma14185322

Academic Editor: Amir Mostafaei

Received: 16 July 2021
Accepted: 10 September 2021
Published: 15 September 2021

Abstract: The great interest, within the fields of research and industry, in enhancing the range and functionality of polymer powders for laser powder bed fusion (LB-PBF-P) increases the need for material modifications. To exploit the full potential of the additivation method of feedstock powders with nanoparticles, the influence of nanoparticles on the LB-PBF process and the material behavior must be understood. In this study, the impact of the quantity and dispersion quality of carbon nanoparticles deposited on polyamide 12 particles is investigated using tensile and cubic specimens manufactured under the same process conditions. The nano-additives are added through dry coating and colloidal deposition. The specimens are analyzed by tensile testing, differential scanning calorimetry, polarized light and electron microscopy, X-ray diffraction, infrared spectroscopy, and micro-computed tomography. The results show that minute amounts (0.005 vol%) of highly dispersed carbon nanoparticles shift the mechanical properties to higher ductility at the expense of tensile strength. Despite changes in crystallinity due to nano-additives, the crystalline phases of polyamide 12 are retained. Layer bonding and part densities strongly depend on the quantity and dispersion quality of the nanoparticles. Nanoparticle loadings for CO_2 laser-operated PBF show only minor changes in material properties, while the potential is greater at lower laser wavelengths.

Keywords: laser powder bed fusion; polyamide 12; nanocomposites; nanoparticles; dispersion; LB-PBF; mechanical properties; additively manufactured parts

1. Introduction

For over 30 years, Additive Manufacturing (AM) has been known for its ability to produce customized parts of high complexity and resolution [1]. Next to the AM of inorganic materials [2,3], laser-based powder bed fusion of organic polymers (LB-PBF-P, according to ISO/ASTM 52911-2:2019) has experienced increased interest in terms of research and development over the last decade [4]. However, AM still requires a lot of operator know-how as many external and internal variables influence the process, e.g., feedstock material [5] or build orientation [6]. In this regard, the interplay of parameters related to the laser source, such as laser power, laser beam diameter, laser scanning speed, and hatch distance, with machine-related parameters, such as powder layer height, process temperature, and recoating speed, is crucial for the successful manufacture of dimensionally accurate three-dimensional parts. The most common way to combine the most important process parameters is the volume energy density [4,7].

Next to the process parameters, first and foremost, bulk solid properties affect the packing density, the flowability and, thus, the spreadability during the AM process [8,9]. If the adaptation of process parameters reaches its limits, the additivation of the base polymer powder with nanomaterials provides a promising tool to steer the material properties in a certain direction or to add new functionalities, e.g., electrical conductivity or magnetism [4,10]. The most common examples of nano-additives for PA12 are carbon-based materials due to their vast availability and attractive properties [4,11,12]. Interestingly, the results of studies on carbon additivation reported in the literature differ significantly for LB-PBF-P. On the one hand, the addition of carbon nanomaterials led to a degradation of mechanical properties under the same process conditions [13–15], while, on the other hand, major improvements were reported [16–19]. However, these improvements mainly correlate with the anisotropic properties of the nanofillers, e.g., carbon nanotubes and fibers, or the optimization of process strategies to improve the processability and the densification of final parts. In this context, a good dispersion of the nanomaterial on the polymer particle is essential to avoid heat accumulation at agglomerate positions and, thus, impairment of mechanical properties [4,15,18]. The quality of the dispersion depends on many influencing factors, for example, the chemical nature of the nano-additives [20], their dosage [21,22], the additivation method [23], and the preparation method of the nanocomposites [15].

While studies have already discussed the importance of a good dispersion for nano-additives on polymer particles [24], a comparative evaluation between different dispersion qualities is still missing, especially for LB-PBF of polymer composites with nanoparticle quantities below 0.1 vol%. Our study closes this gap by processing PA12 powder composites modified with carbon nanoparticles (CNP) [23] into three-dimensional specimens to analyze the influence of two additivation methods of different dispersion qualities on the process and material behavior. In order to gain an initial insight into the influence of another nanoparticle group on the LB-PBF-P process and the mechanical part properties, PA12 specimens with 0.05 vol% colloidally additivated silver nanoparticles (Ag-NP) [20] were manufactured and mechanically tested analogously to PA12/CNP. The results of this study provide a deeper understanding of the importance of the dispersion quality of minute amounts of nanoparticles and its impact on the LB-PBF-P process and the part qualities.

2. Materials and Methods

2.1. Nano-Additivation Process

The adhesion of CNP (CARBON BLACK, Orion Engineered Carbons) onto PA12 powder (EVONIK VESTOSINT 1115, Evonik Industries, Essen, Germany) was achieved by dry coating (DC) and colloidal additivation (Coll) [23,25]. The polymer powder particles are potato-shaped due to their formation through precipitation from ethanol under pressure [26]. The influence of nano-additivation on the polymer shapes is discussed in-depth in [27]. For dry coating, the polymer powder was mixed with CNP powder in a rotating drum for 2 h and then sifted with a 125 µm sieve. For the colloidal deposition process, CNP powder was dispersed in deionized water (washed) by ultrasonic treatment (50 mg/L), followed by laser irradiation with a 10 ps-laser at a wavelength of 532 nm (Edgewave PX400-3-GH, Würselen, Germany, 80 kHz, 30 W, 150 mJ/cm^2, 375 µJ/pulse, 0.25 mm^2 spot size) in a liquid jet setup [25] and finally mixed as a colloid with an aqueous PA12 suspension (50 g/L) [28]. Finally, the colloidal suspension was stirred for 5 min, filtered, dried (24 h at 50 °C), and sifted using a 125 µm sieve. Silver nanoparticles were only added colloidally to PA12 [29,30] for an initial comparison between organic and inorganic nanomaterials, and are discussed in more detail in [20,31].

2.2. Polymer Powder Characterization

Since the size and shape of polymer powder particles have a significant influence on the structure and properties of manufactured parts [32], the different PA12 powder compositions with CNP and Ag-NP were analyzed and characterized by measuring the Hausner ratio, utilizing dynamic image analysis (Camsizer X2, Microtrac Retsch, Haan, Germany) and a ring shear tester (RST-XS, Dietmar Schulze Schüttgutmesstechnik, Wolfenbüttel, Germany) at a pre-consolidation stress of 1 kPa within a small ring shear cell (V = 31 cm^3) [20,23]. An evaluation of the flowability with the Hausner ratio is limited due to its low methodological sensitivity and lack of transferability to the powder application procedure in LB-PBF-P [27]. Therefore, ring shear tests deliver more reliable results, as this method is more sensitive to small changes in powder composition [33,34]. Additional measurements of PA12/CNP and PA12/Ag were performed three times to complete the powder characteristics under the same conditions as previous studies [20,23]. Since only the colloidal additivation was accompanied by a preceding washing step of PA12 powder, comparisons were made with washed and as-received PA12 powder, accordingly. Two-sample *t*-tests were performed to assess the statistical significance.

2.3. Thermal Analysis by Differential Scanning Calorimetry (DSC)

Previous studies analyzed the pure PA12 powder and the additivated powder with CNP and Ag-NP by dynamic DSC (DSC 822e, Mettler Toledo, Columbus, OH, USA) [20,23]. This way, the process temperatures for LB-PBF-P were estimated. After processing the powders in LB-PBF-P (see Section 2.4), cut-up pieces from the center of the manufactured square plates of the different PA12 powder compositions were heated from 25 °C to 230 °C at a rate of 10 K/min to determine the melting temperatures, enthalpies, and crystallinities. The crystallinity X_c of the processed samples was calculated according to Equation (1) [35,36]:

$$X_c = \frac{\Delta H_m}{\Delta H_{100} \cdot \left(1 - w_f\right)} = \frac{\Delta H_m}{209.3\frac{J}{g} \cdot \left(1 - w_f\right)} \tag{1}$$

the heat of fusion of the sample ΔH_m was extracted from the measurements; the heat of fusion ΔH_{100} of 100% crystalline PA12 is found in the literature [37]. The parameter w_f gives the weight percentage of nanoparticles in the composite. Each sample was analyzed three times, leading to a total of 18 runs. The evaluation of the results was performed with the Mettler Toledo STARe Evaluation Software 16.10 (Columbus, OH, USA). The integral tangential baseline was used for the calculation of the relevant enthalpies. Two-sample *t*-tests were performed to include the statistical significance.

2.4. Laser Powder Bed Fusion of Polymers (LB-PBF-P)

Process development of powder bed fusion was conducted with a CO_2 laser-based LB-PBF-P machine (Eosint P385, EOS, Krailling, Germany) of 0.6 mm beam diameter at the working plane. The coater speed was set to 50 mm/s to ensure a smooth and homogeneous powder surface. Pure PA12 powder was used as a base layer of 10 mm height for thermal decoupling from the building platform. To process small amounts of powder, the process chamber was downsized via a reduction in the coater length from 350 mm to 100 mm. In order to avoid premature melting of the top layer of the PA12/CNP powders due to the change in emissivity, the process temperature was lowered successively until coalescence was avoided. Conversely, the PA12/Ag powders needed to be processed at higher temperatures. Furthermore, the power of the IR-emitters was set to the same power range in the machine software for processing of pure PA12 in order to provide similar thermal conditions for every material composition. A powder layer of 0.6 mm was added to the last molten areas to reduce warpage during the cooling phase. Five tensile 1BA specimens (DIN EN ISO 527-2) of 2 mm thickness and five square plates ($15 \times 15 \times 2$ mm^3) were produced in the x–y-plane with a layer height of 100 µm for each material composition. An alternating hatching was chosen for every layer between the x and y directions. The aim was to find a set of process parameters that could be used to process every material composition under the same conditions instead of optimizing the process parameters towards the best processability, density, or mechanical properties for individual material composites. In this way, the influence of nanoparticles on the material behavior and the material limitations of the composites could be examined and evaluated. However, the same energy density could not be used for 0.05 vol% CNP. Lowering the energy density of the reference material of pure PA12 was not an option as it led to a lack of layer bonding. Thus, the laser power was lowered successively to such an extent that the thermal difference between the material temperature and the laser energy input was reduced, and curling no longer occurred. To evaluate the performance of the integrated CO_2 laser source, laser power measurements were conducted with a laser power meter (LM-200, Coherent, Santa Clara, CA, USA). In the onboard software of the LB-PBF machine, the percentage of the laser power was set in 1% steps from 5% to 10% and in 5% steps from 10% to 100% laser power. In doing so, the nominal and real output power values could be extracted for the experiments (Figure S1). Table S1 summarizes the process parameters and resulting energy densities for the different powder composite materials. At the end of the building process, the IR-emitters were switched off. The pyrometer data of the LB-PBF machine showed that the powder bed temperature dropped from the process temperature to 120 °C, which is below the endset temperature of crystallization, at a declining cooling rate of approximately 15 K/min to 1 K/min. The process chamber was allowed to further cool down and remained closed for at least 12 h.

2.5. *Dimensional Accuracy and Tensile Testing of Specimens*

The thickness and the width of the measuring range of the tensile bars were measured with a micrometer with an accuracy of 0.01 mm. Tensile tests were performed according to DIN EN ISO 527 on the universal testing machine (Quasar 100, Cesare Galdabini, Vigevano, Italy) at room temperature with a load cell of 10 kN and a contact extensometer. The Young's modulus was measured at a speed of 0.5 mm/min until an elongation of 0.3% was reached, followed by a speed of 20 mm/min to measure the ultimate tensile strength and the ultimate elongation.

2.6. *Polarized Light and Scanning Electron Microscopy (SEM)*

The square LB-PBF specimens were cut with a microtome (Leica Biosystems, Wetzlar, Germany) to 10 µm slices and placed on microscope slides with immersion oil. A light microscope (Metalloplan, Leitz/Leica, Wetzlar, Germany) with two polarizers was used to analyze the layer bonding and the birefringence of the crystalline structures. The surfaces and tensile fractured surfaces of PBF specimens were imaged using a scanning electron microscope (GeminiSEM 500, Zeiss, Oberkochen, Germany) equipped with an SE2 detector, an acceleration voltage of 1 kV, and an aperture of 15 µm. Images were taken of the top surface, as well as the edges and centers of the fractured surfaces of the tensile specimens.

2.7. *X-ray Diffraction (XRD) and Infrared (IR) Spectroscopy*

X-ray diffraction (Empyrean series 2, Malvern Panalytical, Worcestershire, UK) using Cu K_α radiation (154 pm, 40 kV, 40 mA) was performed with an Empyrean diffractometer (Panalytical) in Bragg–Brentano geometry. The incidence beam optics comprised the Bragg–Brentano-HD module, fixed divergence (1/8°), anti-scatter (1/8°) slits, and 0.04 rad Soller slits. The diffracted beam optics comprised Soller slits (0.04 rad), an anti-scatter slit (16.8 mm), and a GaliPIX 3D detector (Panalytical). Sample specimens were fixed in the sample holder and rotated during measurement. The diffractograms were collected in the range of $5° \leq 2\theta \leq 90°$ at a step size of 0.014° and a measuring time of 25 s per step. Phase analysis and Rietveld refinements of the obtained diffraction patterns were performed with Profex 4.3 (Solothurn, Switzerland), a GUI of the BGMN Rietveld Analysis Program (Dresden, Germany), to determine the phase composition and the unit cell parameters, but also to quantify the crystallite sizes. Infrared (IR) spectra of the specimens were recorded in attenuated total reflection (ATR) geometry in the spectral range from 6000 cm^{-1} to 400 cm^{-1} at a resolution of 2 cm^{-1} using a Fourier-transform infrared (FT-IR) spectrometer (FTS3100, Digilab, Hopkinton, MA, USA). The spectra were normalized to their respective maximum absorbance. Baseline corrections were performed manually. Pure PA12 and PA12/CNP specimens were analyzed this way.

2.8. Micro-Computed Tomography (μ-CT)

The specimens of PA12 and PA12/CNP were examined by X-ray microtomography to determine the process-induced porosity and the pores' sphericity. This enabled the analysis of pore size and morphology in the micrometer range and the statistical evaluation of the pore characteristics. The type XT H 160 μ-CT system (Nikon, Tokyo, Japan), equipped with a microfocus X-ray source (tungsten) with a maximum voltage of 160 kV and a 3 μm threshold for the 3D scan, was used for the investigations. The 2D images were reconstructed into a 3D image using the CT Pro 3D software (Nikon, Tokyo, Japan). Subsequently, it was loaded into the analyzing and visualizing software VGStudio Max 2.2 (Volume Graphics, Heidelberg, Germany). The corresponding scanning parameters can be seen in Table 1.

Table 1. Scanning parameters for the computed tomography scans (μ-CT).

Material	Beam Energy	Beam Current	Power	Effective Pixel Size	Exposure Rates
PA12/ PA12-CNP	99 kV	26 μA	2.5 W	15 μm	1.42 s, 0.707 fps

After the scan, the images were reconstructed and loaded into the VGStudio Max 2.2 analysis and visualization software (Volume Graphics GmbH, Heidelberg, Germany). Afterwards, algorithm-specific voxels were characterized with the "VGDefX (v2.2)" as defects based on their grey value compared to a defined local threshold for contrast. The theoretical resolution of the system is 3 μm; however, due to the dimensions of the specimens, the minimum accurate detectable pore size increased to 15 μm. By setting the minimum pore size in the defect analysis program according to the effective pixel size of the specimens (15 μm) and checking the detected probability of the detected pores, the pores could be distinguished from the noise.

3. Results and Discussion

3.1. Material Characterization of Polymer Powder Composites

An overview of all measured powder characteristics can be found in Table S2. While the difference between the flowability values measured with the Hausner ratio is insignificant, the ring shear test results indicate distinguishable deviations. Figure 1 summarizes the flowability (ffc-value) results of the differently additivated PA12 powders.

For instance, 0.005 vol% of CNP significantly increases the flowability of PA12 powder (*; $p \leq 0.05$) when additivated with the colloidal approach, while dry coating significantly (****; $p \leq 0.0001$) lowers the free-flowing (ffc > 10) powder properties to easy-flowing (ffc < 10) [38]. An increase in the CNP dosage to 0.05 vol% does not impair the flowability by colloidal additivation, while dry coating further reduces the flowability of the powder. The poorer outcomes can be ascribed to the stronger mechanical forces during the dry coating process, which lead to an increased inter-particle cohesion due to fines [23,27]. Regardless of the additivation method, 0.05 vol% of CNP leads to poorer flowing powders than smaller quantities due to an increased amount of CNP agglomerates on the polymer particle surfaces [23]. In comparison, the colloidal additivation of the same amount of

Ag-NP slightly increases (ns; $p > 0.05$) the flowability. Based on these results, the colloidal additivation process is the more suitable option for the additivation of nanoparticles since the good flowability of the base powder material is maintained.

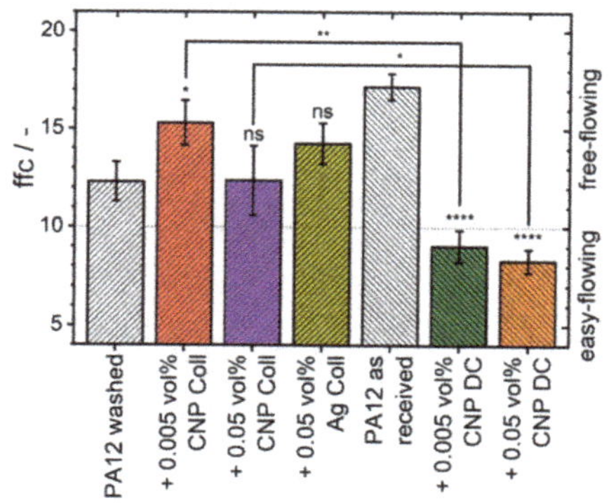

Figure 1. Flowability characteristics of PA12 powder and its composites measured by a ring shear cell at a pre-consolidation stress of 1 kPa. Due to the additional washing step before the colloidal additivation, colloidal composite powders are compared to washed PA12 powder while dry-coated composites are compared to PA12 powder as received. An increase in significance is depicted with an increase in the number of asterisks, while no significant differences are declared as "ns" ($p > 0.05$). Results are based on three measurements.

3.2. *Thermal Evaluation by Differential Scanning Calorimetry (DSC)*

Evaluating the melting behavior of the PA12 powders during the heating stage of the DSC helps to estimate the processing temperature for LB-PBF-P. Regardless of the nano-additives, each powder composition exhibits an endothermal increase in heat flow at around 170 °C during the heating phase. This value sets the starting point for the LB-PBF-P experiments. The processed samples show a second peak around 190 °C, which correlates with the higher melting temperatures of unmolten particles [39]. This peak becomes more pronounced as more CNP nanoparticles are added (Figure 2), which means that the number of unmelted particles increases with the dosage of nanoparticles.

Consequently, it provides an initial indication of the poorer expected material properties of the composites compared to pure PA12 due to reduced part density and layer bonding. However, the additions of 0.005 vol% CNP and 0.05 vol% Ag-NP retain the curve characteristics of pure PA12. The exact dimensions of the second peak of every material were not averaged due to large deviations between the three runs of each composition (Figure 2). Thus, the quantitative difference between the material compositions and the additivation methods is insignificant.

Since the second peak does not change the main peak positions, the temperature onset, peak, and endset do not change significantly (ns; $p > 0.05$) by any additivation (Figure 3a). Yet, the overall crystallinity increases significantly for all additivated samples except for PA12 colloidally additivated with 0.005 vol% CNP (Figure 3c) due to the broadening of the second peak. The differences between the two additivation methods are more prominent for 0.005 vol% CNP (**; $p \leq 0.01$) than for 0.05 vol% CNP (*; $p \leq 0.05$). Therefore, an increase in crystallinity is not an initial indication of improved mechanical properties since the higher crystallinity values may come from an increased number of unmelted, but highly crystalline, polymer particles. Moreover, the results of the heat of fusion and crystallinity do not provide information about layer bonding. If an increase in crystallinity by nanoparticles has macroscopic advantages for the properties of the mechanical part, further analytic examinations are required. For instance, microscopic investigations need to be conducted to evaluate the connection between subsequent layers and the crystal growth behavior at these interfaces.

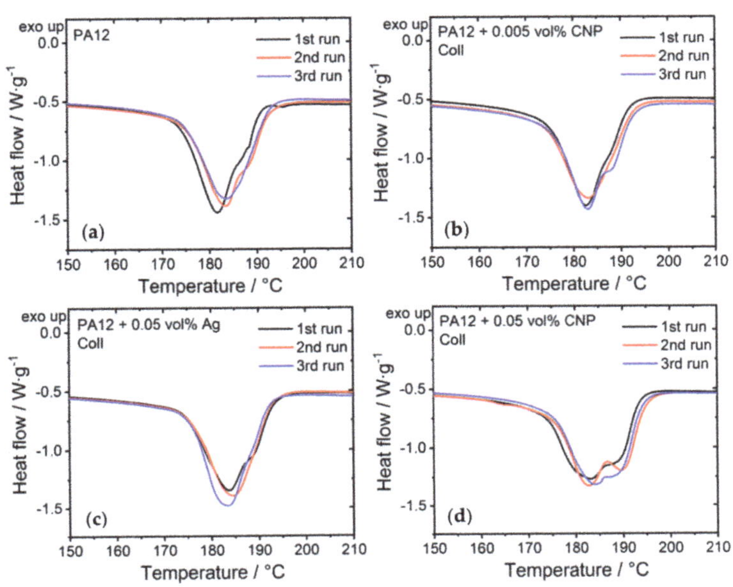

Figure 2. Three heating curves of the LB-PBF-P specimens made of (**a**) pure PA12, (**b**) PA12 and 0.005 vol% carbon nanoparticles, (**c**) PA12 and 0.05 vol% silver nanoparticles, and (**d**) PA12 and 0.05 vol% carbon nanoparticles. The addition of the nanoparticles was performed via colloidal additivation. The heating rate was 10 K/min.

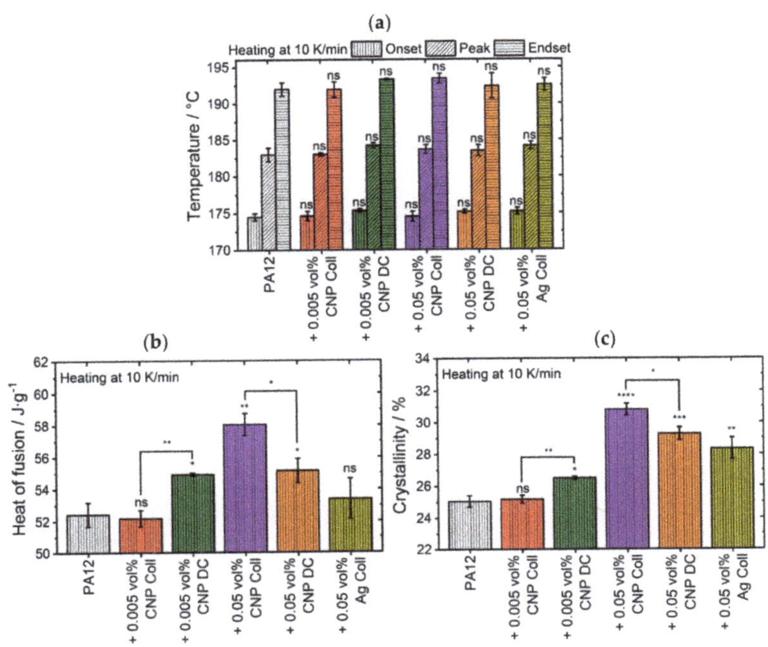

Figure 3. Averaged thermal values of different material compositions showing their results of (**a**) peak melting temperature values, (**b**) heat of fusion, and (**c**) crystallinity. The level of significance increases with the number of asterisks, while "ns" stands for an insignificant difference.

3.3. LB-PBF of PA12 and Composites

Differences between colloidal additivation and dry coating were noticeable during powder recoating, where the colloidally additivated powder could be spread more homogeneously than the dry-coated powder. During processing, specimens of colloidally additivated PA12 powder of 0.05 vol% CNP required less energy (0.115 J/mm^3) to be built compared to the dry-coated variant (0.127 J/mm^3). The temperature had to be decreased from 171 °C to 169 °C when increasing the CNP amount to 0.05 vol%. Otherwise, overly high temperatures and energy densities lead to a displacement of the specimens' molten layers, ultimately resulting in a build job failure. Thus, lower temperatures and less energy were required to process and bond the polymer layers with the addition of minute amounts of CNP. By contrast, PA12 with 0.05 vol% Ag could be processed without any curling after raising the temperature by one degree Celsius while maintaining the same process parameters and energy density as pure PA12. When qualitatively comparing the spreadability of CNP and Ag-NP at 0.05 vol% during the recoating process, the powder bed of additivated PA12 powders with Ag-NP appeared more homogeneously distributed than the CNP counterpart. At a lower dose of 0.005 vol%, the dry-coated CNP powders exhibited faster curling after powder application than the colloidally additivated ones. It was found that the reduced process window of CNP additivated powders [23] leads to faster warpage of specimens during the cooling phase.

3.4. Microscopic Evaluation of LB-PBF-P Samples

Polarization images at low magnification (Figure 4) reveal the quality of the layer bonds of each processed material composition. The presented images show horizontal layers built in the z-direction. Pure PA12 shows consistent bonding of individual layers (Figure 4a) with some unmelted particles, which is a typical phenomenon for LB-PBF-P results. The degree of particle melt could be increased by higher laser energy inputs, but the aim of this study was to evaluate the influence of nanoparticles on the processability and material properties. The first indications of a reduction in particle melting became apparent when adding 0.005 vol% CNP to the surfaces of the PA12 particles. Even though the process temperature remained the same, layer bonding was partially disrupted by an increased number of agglomerated unmelted particles (Figure 4b,c). Visual differences between the colloidal and dry-coated additivation procedure were non-existent at this low volume of CNP.

When increasing the dosage of CNP to 0.05 vol%, the laser power had to be lowered by 15% for dry-coated powders and 24% for colloidally additivated powders (Table S1) to process 20 layers without build failures. However, this led to a deterioration in bonding quality between the layers. Although the dry-coated specimens absorbed an energy of 0.127 J/mm^3, their microscopic results show larger and longer gaps between the layers (Figure 4e) than the colloidal additivated specimens (Figure 4f), which received a lower laser energy of 0.115 J/mm^3. More agglomerated CNP from dry coating are most likely the cause of this outcome. By comparison, good layer bonding is given for the same amount of Ag-NP (Figure 4d) with occasional smaller gaps. Reasons for the differences between CNP and Ag-NP are either due to the energy distribution, which is redirected by the absorptive and emissive properties of the CNP, or the polymer–polymer connection, which is interrupted by nanoparticle interactions of opposite layers. Another reason could be the difference in the Hamaker constant between CNP and Ag-NP [40,41]. In addition, the change of bonding behavior between the polymer and the organic CNP or inorganic Ag-NP could be the cause.

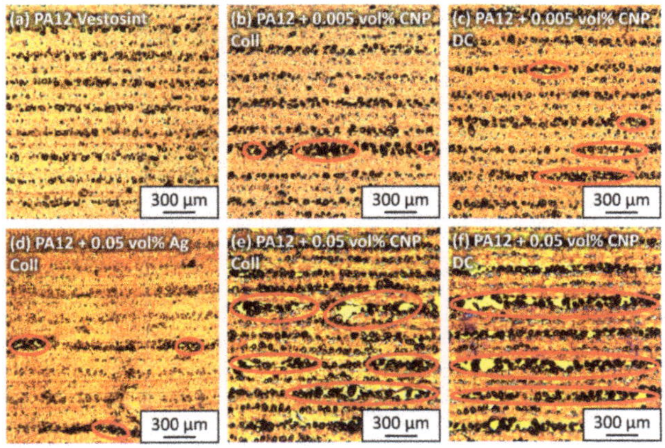

Figure 4. Microscopic polarized images of 10 µm sliced LB-PBF-P specimens made of (**a**) pure PA12, (**b**,**c**) PA12 and 0.005 vol% carbon nanoparticles, (**d**) PA12 and 0.05 vol% silver nanoparticles, and (**e**,**f**) PA12 and 0.05 vol% carbon nanoparticles. PA12 powders were additivated (**b**,**d**,**e**) with the colloidal deposition and (**c**,**f**) with the dry coating method. The images provide an overview of the processed layers in a horizontal position, where higher amounts of carbon nanoparticles lead to poorer layer bonding.

With microscopy images of higher magnification (Figure 5), lamellar structures can be identified. The typical crystalline structures of PA12 with lamellae up to 25 µm are visible, originating from unmelted particle cores or random impurities (Figure 5a). The dimensions of these lamellae, which develop during the cooling phase of LB-PBF, correlate well with the microscopic results of the calorimetric powder measurements [23] since their lengths lie between the values of the cooling rates at 0.5 K/min and 20 K/min. Spherical and ellipsoidal structures can be found, as observed in previous studies [23,25]. Adding 0.005 vol% CNP does not increase the frequency of crystalline structures, but seemingly reduces them. There are no apparent differences between the two additivation methods at this low dose of nanoparticles (Figure 5b,c). However, formations of agglomerated nanoparticles to chain-like structures (Figure 5e,f) can be identified at 0.05 vol% CNP. Even though the nanoparticles occasionally induce lamellar growth across layer boundaries, they also introduce new interfaces that predetermine mechanical weak points. While there are larger CNP clusters in the dry-coated specimens, the colloidal additivation leads to a more homogeneous distribution of the CNP in the melt. It is expected that dry-coated samples will have poorer mechanical properties due to the higher frequency of clusters. The lamellar structures cannot be identified in the PA12 sample with 0.05 vol% Ag-NP (Figure 5d). Only crystalline growth from unmelted particles can be found. As expected from the thermal powder analyses [20], no nucleation effects of the Ag-NP can be detected by microscopic evaluations. However, both microscopic and DSC evaluations are limited as they only represent a small percentage of the total sample. Three-dimensional analysis methods, such as µ-CT (see Section 3.8), can help to better understand the internal structures of the entire sample.

Our previous hypothesis [23] that CNP can introduce a certain anisotropy into the material system, which can be beneficial for increasing the layer bonding, cannot be confirmed. Instead, a CNP dose of 0.05 vol% is high enough to compromise layer bonding if the process parameters are not changed. However, lower quantities can be used to tailor mechanical properties.

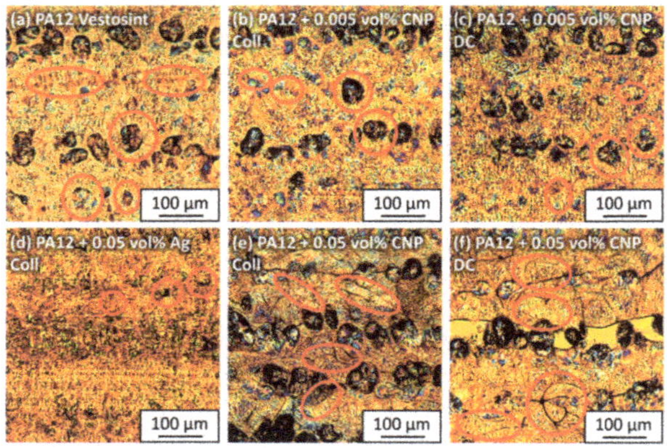

Figure 5. Magnified microscopic polarized images of 10 µm sliced LB-PBF-P specimens made of (**a**) pure PA12, (**b**,**c**) PA12 and 0.005 vol% carbon nanoparticles, (**d**) PA12 and 0.05 vol% silver nanoparticles, and (**e**,**f**) PA12 and 0.05 vol% carbon nanoparticles. PA12 powders were additivated (**b**,**d**,**e**) with the colloidal deposition and (**c**,**f**) with the dry coating method. The images provide a more detailed view of the developed crystalline structures and the positions of carbon and silver nanoparticles in the cooled polymer melt.

3.5. Dimensions and Mechanical Properties of Tensile Bar Specimens

The dimensions of the reference specimens of pure PA12 were 2.01 ± 0.02 mm in thickness and 4.66 ± 0.02 mm in width (Figure 6a). The specimens exhibit a Young's modulus E of 1.93 ± 0.07 GPa, an ultimate tensile strength σ_{ult} of 46.4 ± 0.5 MPa, and an ultimate elongation ε_{ult} of 5.2 ± 0.2% (Figure 6b,c). These values are well within the typical range of mechanical properties of specimens produced by LB-PBF-P [4]. The exact target width of 5 mm was not achieved since no contour exposure was used in this study. However, since the values are within the range of the DIN standard and show only a small spread, the process parameters still provide a good reproducibility for pure PA12 powders in LB-PBF.

The low deviations among the five specimens prove the consistently good processability of the powder composites (Figure 6), especially at low doses of nanoparticles (<0.05 vol%). However, the deviations increase at higher nanoparticle concentrations, which corresponds to the poorer processabilities of these powder composites. Further differences are caused by the additivation methods. When comparing the dry-coated PA12 containing 0.005 vol% CNP with pure PA12 (Figure 6a), the width of the dry-coated specimens is significantly (0.9%) (****; $p \leq 0.0001$) lower, while the thickness values are significantly (20.7%) (****; $p \leq 0.0001$) higher. Faster warpage of dry-coated specimens with 0.005 vol% CNP during cooling lead to the specimens' greater thickness and narrower width. This results from the reduced processing window due to an increase in crystallization temperatures induced by the nucleation effect of CNP [23]. By comparison, the colloidally additivated specimens of the same CNP concentration exhibit a significant (****; $p \leq 0.0001$) increase in thickness and width by only 1.2% and 7.8%, respectively. Since there are no significant thermal differences between these two 0.005 vol% PA12/CNP powders, the main reason should be ascribed to the quality of the nanoparticle deposition and an increased expected agglomeration of nanoparticles in the melt. Overall, the specimens built from composites show significantly higher dimensional values than those from pure PA12. This correlates with the unintentional melting and coalescence of neighboring particles induced by the CNP (see Section 3.6).

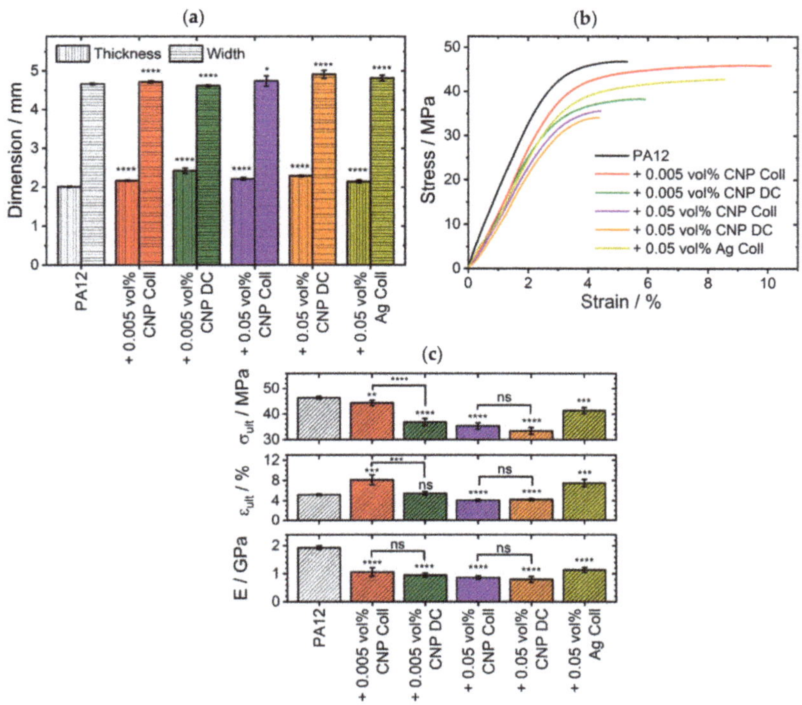

Figure 6. The overview of the tensile test results shows the LB-PBF-P specimens of different material compositions with regard to their (**a**) measured dimensions, (**b**) exemplary stress–strain curves, and (**c**) mechanical properties of the ultimate tensile strength (σ_{ult}), ultimate elongation (ε_{ult}) and Young's modulus (E). The level of significance increases with the number of asterisks, while "ns" stands for an insignificant difference.

The difference between the two additivation methods at minute amounts of 0.005 vol% CNP can also be recognized in the tensile results (Figure 6b,c). Here, dry-coated specimens exhibit a σ_{ult} that is significantly (20.6%) (****; $p \leq 0.0001$) lower than pure PA12, and a ε_{ult}, which increases insignificantly (5.4%) (ns; $p > 0.05$). In comparison, the colloidal additivation of the same quantity of CNP also decreases σ_{ult} significantly (**; $p \leq 0.01$), but only by 4.4%, while ε_{ult} is increased significantly (***; $p \leq 0.001$) by 56.5%. While the difference between the dimensional accuracy (Figure 6a) and the mechanical properties (Figure 6b,c) diminishes when increasing the volume of CNP to 0.05 vol%, there are small but insignificant (ns; $p > 0.05$) differences between colloidal additivation and dry coating. For instance, the width increases significantly (****; $p \leq 0.0001$), by 5.4%, and the thickness increases significantly (****; $p \leq 0.0001$), by 14.1%, for dry-coated specimens. On the other hand, the dimensions of the colloidally additivated specimens have a smaller, but significant (*; $p \leq 0.05$), growth of 1.7% and 10.2% (****; $p \leq 0.0001$). Similarly, every measurand of the mechanical properties is significantly (****; $p \leq 0.0001$) deteriorated at 0.05 vol% of CNP. Interestingly, the colloidal additivation showed a 4.9% higher tensile strength and a 3.3% higher Young's modulus than the dry-coated specimens, even though the colloidally additivated powder received 9.4% less energy during the LB-PBF-P process (Table S1). However, this difference between the two additivation methods is insignificant (ns; $p > 0.05$). When changing the nano-additives to 0.05 vol% silver, the dimensions increase significantly (****; $p \leq 0.0001$) by 3.3% in width and 6.9% in thickness. At the same time, the mechanical properties are close to values of 0.005 vol% CNP by colloidal additivation. Finally, all composites have a significantly (****; $p \leq 0.0001$) lower

Young's modulus than pure PA12 (Figure 6c). The reasons for the decrease in tensile strength and Young's modulus, but the increase in elongation, could be due to a weakened interaction between the nanofiller and the polymer matrix (e.g., low crosslink density) [13]. A low dispersion and, thus, a higher probability of agglomeration of nanoparticles would explain the overall impairment of dimensions and mechanical properties by dry coating of 0.005 vol% CNP. Conversely, the chances of nanoparticles agglomerating increase at higher quantities of CNP, limiting the possibility of a better dispersion to induce macroscopic changes. Other nucleating nanomaterials should be chosen instead of CNP to improve the interaction between the nano-additives and the polymer matrix.

In general, we could show that small volume fractions of CNP and Ag-NP are already enough to significantly influence the dimensional and mechanical properties of LB-PBF specimens. The quality of the dispersion plays a decisive role in this. The impact of CNP on the dimensional and mechanical properties is greater than that of the same dose of Ag-NP without nucleation properties. However, for CO_2 laser-operated powder bed fusion processes, the influence of nanoparticles on the material behavior is limited and usually does not contribute to better mechanical properties. In this case, a compromise has to be made between introducing new material properties through nanoparticles (e.g., plasmonic [31,42] or magnetic properties [10]) and the mechanical properties of additively manufactured parts. However, if the wavelength of the laser source is in the near-infrared or visual wavelength range, absorption-enhancing nanoparticles of high dispersion become inevitable to effectively process polymer powders at these wavelengths [31].

3.6. *Tensile Fractography*

Images of the top surface of the specimens help to evaluate the quality of the molten state of the outer layers, while images of the fractured surfaces are used to identify the fracture behavior under the influence of different amounts of nanoparticles. The presented images depict the processed horizontal layers perpendicular to the build direction (Figure 7).

PA12 shows overall good layer bonding without visible layer boundaries in the body of the sample (Figure 7b), while some lack of bonding exists around the edges (Figure 7a). The reason for this is the absence of contour parameters in this study, which are usually used in LB-PBF-P to mitigate this phenomenon. The layer bonding and, thus, the mechanical properties can be further improved by increasing the degree of particle melt with higher laser energy densities. However, this usually sacrifices the parts' resolution and dimensional accuracy due to unwanted sintering of adjacent loose powder particles [26]. The fracture surface images primarily reveal brittle regions with some ductile areas around the edges of the sample (Figure 8a), which are believed to be the origin of fracture during tensile testing [26]. Upon closer inspection, the ductile areas consist of mainly spherical fibrillated structures indicating broken particle cores (Figure 8b). The condition of the top surfaces (Figure 7c) is typical for processed PA12 powder with partially molten and unmolten particles from the surrounding powder due to bleeding of thermal energy [43]. However, this has no adverse effect on the dimensional accuracy of the specimens since the thickness of the final parts is at the target value of 2 mm (Figure 6a). The seamless bonding of the layers further supports the good dimensional accuracy of the PA12 specimens. Optionally, the surface quality could be improved by an additional post-processing step involving grinding or polishing [44].

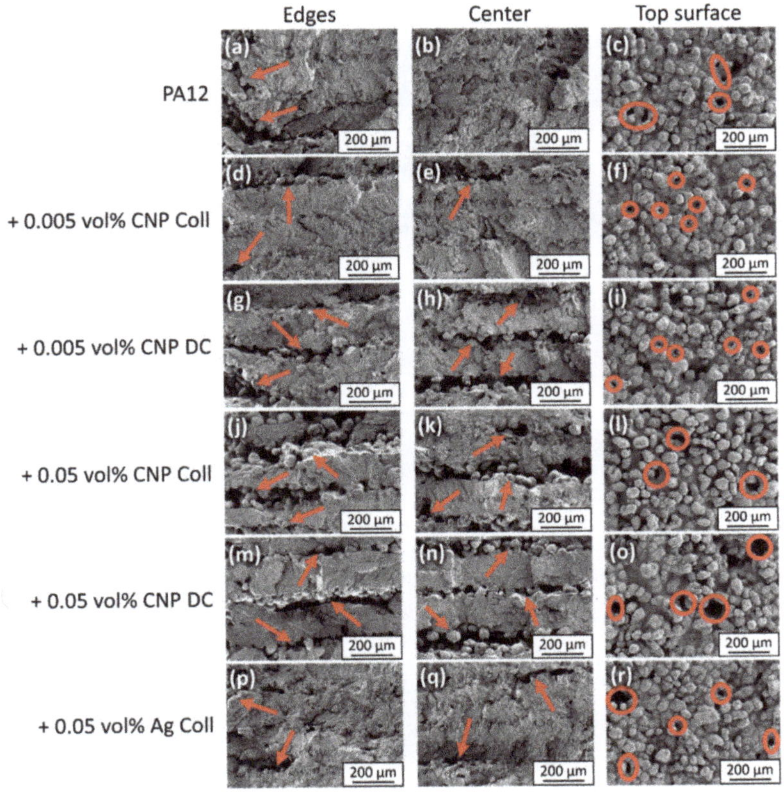

Figure 7. Collection of scanning electron microscopy images of the fractured surfaces of the tensile bars and of the top surface of the specimens. The left column (**a**,**d**,**g**,**j**,**m**,**p**) depicts the edges, and the middle column (**b**,**e**,**h**,**k**,**n**,**q**) shows the center of the fractured surface. The condition of the specimens' top surface can be seen in the right column (**c**,**f**,**i**,**l**,**o**,**r**). The quantity of nanoparticles increases from top to bottom. Exemplarily, the red arrows mark unmelted polymer particles, while the red circles highlight voids.

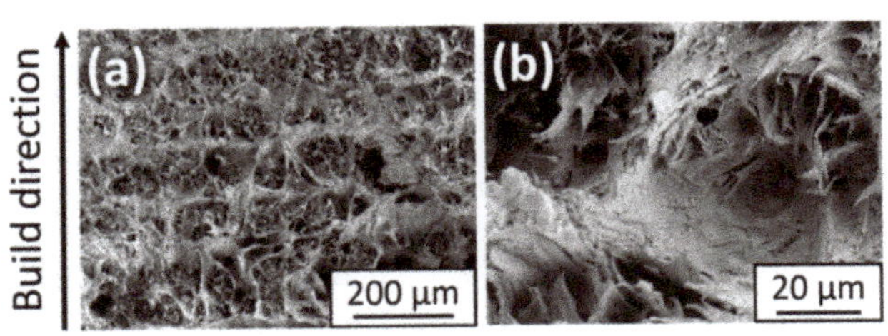

Figure 8. (**a**) Scanning electron microscopy images of the ductile area of fractured surfaces of the PA12 specimens. A more detailed view of the spherical fibrillated structures responsible for the ductility can be seen in (**b**).

Adding 0.005 vol% CNP by colloidal additivation leads to more distinct layer boundaries of partially unmelted particles near the edges and in the center of the specimens (Figure 7d,e), while interlayer bonding is still present. The same quantity of CNP, but additivated by dry coating, results in enlarged gaps between individual layers near the edge and inside of the specimens (Figure 7g,h). These interrupted transitions are due to partially or fully unmelted particles, which correlate with the second peak, at around 190 °C, observed in DSC (Figure 2). The frequency of ductile areas is the highest for PA12 with 0.005 vol% CNP by colloidal additivation compared to pure PA12 and other specimens. This can be ascribed to the increased ultimate elongation of the tensile specimens (Figure 6c). Due to the same amount of CNP being present, the reasons for these differences are either the quality of dispersion [23] or the change of powder flowability during additivation, which influence the processability of the composite powders. These apparent differences in fractured surfaces between the colloidal additivation and the dry coating of 0.005 vol% CNP are reflected in their dimensional accuracy (Figure 6a) and mechanical properties (Figure 6b,c). The higher number of gaps explains the significant increase of 20.7% in thickness for the dry-coated specimens. However, the top surface structures show no evident differences between the two additivation methods for the specimens additivated with 0.005 vol% CNP (Figure 7f,i) and are equivalent to the pure PA12 results. The difference in fracture behavior between the two additivation methods is undistinguishable when increasing the dosage of CNP to 0.05 vol% (Figure 7j,k,m,n). These findings coincide with the insignificant differences (ns; $p > 0.05$) in mechanical properties between the two differently additivated specimens (Figure 6c). The number of unmelted particles between the layers increases, further impairing the mechanical properties of the specimens. The only noticeable difference in the top surface structure is a higher number of voids within the molten surface. These voids are more pronounced for the dry-coated specimens than for the colloidally additivated ones. This could be an indication of an increased amount of escaping gas or polymer chain scission.

By contrast, the same quantity of Ag-NP leads to very similar results as pure PA12 with regard to layer bonding and layer boundary conditions (Figure 7p,q). Since Ag-NP do not induce crystal growth, in contrast to CNP, under LB-PBF cooling conditions (Figure 5d) [20], they can maintain the good processability of pure PA12 powder. Finally, a dispersion of good quality is particularly important for nucleating nanoparticles below a dose of 0.05 vol% if a shift in mechanical properties is desired.

3.7. X-ray Diffraction (XRD) and Infrared (IR) Spectroscopy

Polyamide 12 is known to crystallize depending on the acting stresses, temperature, and pressure in different polymorphic forms. Four crystalline phases, namely the alpha (α), alpha' (α'), gamma (γ) and gamma' (γ') phases, are known [26,45–47], with the gamma phase being the most stable form at ambient conditions and the alpha form observed for PA12 annealed at elevated pressures [48]. In addition, an intermediate α'' form was observed prior to the transformation of γ PA12 to the γ' polymorph by drawing [49], respectively, the (intermediate) crystallization of PA12 from the melt in the α' form with a subsequent transition to the γ form while cooling to room temperature [50]. The γ' phase can be produced by melt quenching [47]. Precipitated PA12 powders have been reported in the literature to frequently show an 'intermediate' structure between the alpha and gamma phase that is characterized by two distinct reflexes found at 2θ around 20.9° and 22.0° (Cu K_{α}) [23,26,50].

Diffraction patterns of the specimen produced from the different composite powders in comparison to PA12 powder are depicted in Figure 9 below. The experimental pattern could be described solely by the presence of a crystalline monoclinic (pseudohexagonal) γ PA12 phase when taking the structural data reported by Cojazzi et al. [51] and a polynomial background function into account, i.e., there are no indications of the presence of another crystalline PA12 phase. The specimens were produced by LB-PBF-P, i.e., the powder was first melted and then allowed to cool at relatively moderate cooling rates. At these conditions, i.e., the crystallization of PA12 from the melt, the γ form was previously reported to be formed [50]; however, also in LB-PBF-produced specimens, an intermediate structure between the α and γ phase was recently reported to be present in the precipitated feedstock [26]. Consequently, despite an apparent shift of the main reflex of less than 1° (Figure 9b), we can also confirm that under the chosen conditions, the CNP present in the composite powder do not trigger the nucleation of PA12 polymorphs other than γ in the built specimens. The unit cell parameters for the γ PA12 phase deduced by Rietveld refinement of the experimental diffraction pattern are summarized in Table 2. No trend indicating a dependence of unit-cell parameters or lattice spacings (Table S3) on any amount of NP present in the powder system could be deduced.

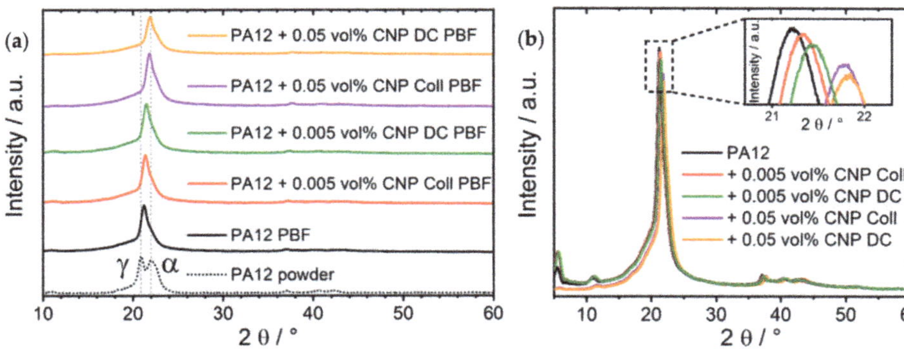

Figure 9. Diffraction patterns of PA12 powder and specimens of different compositions additivated by colloidal deposition and dry coating, depicted (**a**) as overviews separated from each other and (**b**) on top of each other with a zoomed-in picture of the shifted main reflex positions.

Table 2. Unit cell parameters for the monoclinic gamma PA12 phase as determined from Rietveld refinements of the diffraction pattern of processed specimens.

Material Composition	a in nm	b in nm	c in nm	β in °
PA12	0.4838	3.1810	0.9484	121.2
PA12 + 0.005 vol% CNP Coll	0.4838	3.1912	0.9513	120.8
PA12 + 0.05 vol% CNP Coll	0.4838	3.2219	0.9484	121.2
PA12 + 0.005 vol% CNP DC	0.4838	3.1854	0.9498	121.2
PA12 + 0.05 vol% CNP DC	0.4838	3.2213	0.9484	121.2

The crystallite sizes of the γ PA12 phase in the composite specimens were exemplarily determined from the refinement of the diffractograms for the (100), (002), and (020) orientations, characterized by 2θ angles of 21.4°, 21.9° and 5.5°, respectively (Table 3). No preferred crystallite orientation or texture can be deduced from the diffractograms. With increasing carbon black content, a slight increase in crystallite size in the (001) and (002) orientation can be noted, although this effect is relatively small.

Table 3. Crystallite sizes in 100, 002, and 020 orientation for gamma PA12 in composite powder specimens as determined from Rietveld refinement.

Material Composition	Crystallite Size in nm			Error Crystallite Size in nm		
	(100)	(002)	(020)	(100)	(002)	(020)
PA12	8.21	7.12	6.478	0.15	0.14	0.099
PA12 + 0.005 vol% CNP Coll	8.53	6.58	6.426	0.16	0.14	0.041
PA12 + 0.05 vol% CNP Coll	9.43	8.66	5.503	0.19	0.11	0.057
PA12 + 0.005 vol% CNP DC	6.56	5.92	6.691	0.09	0.14	0.051
PA12 + 0.05 vol% CNP DC	9.29	9.51	5.037	0.20	0.14	0.055

According to Bain et al. [26] and Rhee and White [52], respectively, γ PA12 can be discriminated from α PA12 by the position of characteristic vibrations in the IR range, e.g., the Amide I (1635 cm^{-1} (α) vs. 1640 cm^{-1} (γ)) or the Amide II band (1540 cm^{-1} (α) vs.

1563 cm^{-1} (γ)). The band positions for amide I and amide II observed in ATR spectra of the samples collected at a resolution of 2 cm^{-1} are summarized in Table 4. Because of the instrumental resolution and the experimentally observed band positions, e.g., for the amide I between 1635 cm^{-1} and 1638 cm^{-1}, respectively, representing very weak amide II bands, a clear assignment to one of the crystal phases or the deduction of the presence of a crystal mixture cannot be made from the IR spectra.

Table 4. Positions of the amide I and amide II bands in PA12/CNP composite specimens.

Material Composition	Amide I in cm^{-1}	Amide II in cm^{-1}
PA12	1637	1543 and 1566
PA12 + 0.005 vol% CNP Coll	1637	1541
PA12 + 0.05 vol% CNP Coll	1635	1547 and 1566
PA12 + 0.005 vol% CNP DC	1638	-
PA12 + 0.05 vol% CNP DC	1636	1545

3.8. Micro-Computed Tomography (μ-CT)

The effect of the additivation of CNP on the relative density of PA12 specimens is visible in Figure 10. Relatively large line-shaped defects in the PA12 specimens additivated with 0.05 vol% CNP and their layer-wise arrangement perpendicular to the building direction confirm the insufficient diffusion between the scanning layers. However, the defects are more distinguishable in the dry-coated specimens of 0.05 vol% CNP, where the lack of diffusion is more visible at the mid-layers of the specimens (Figure 10c,d).

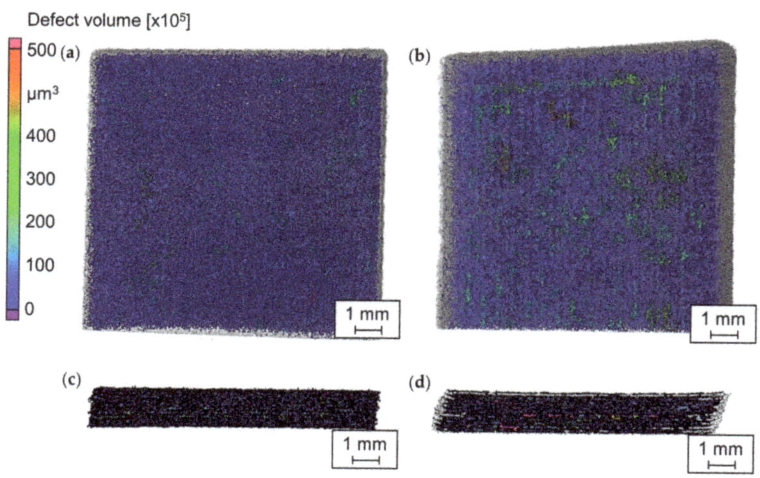

Figure 10. Results of μ-CT scans for specimens of pure PA12 in (**a**) top and (**c**) side view and of PA12 dry-coated with 0.05 vol% CNP in (**b**) top and (**d**) side view.

It can be seen in Figure 11a that pure PA12 has the highest relative density of 89.0%. The addition of 0.005 vol% CNP reduces their relative density to 88.7% and 88.6% for the dry coating and the colloidal additivation methods, respectively. An increase in CNP to 0.05 vol% further decreases the relative density of the PA12 specimens to 83.3% and 84.1% for the dry coating and the colloidal additivation methods, respectively. This leaves comparable densities at 0.005 vol% CNP between dry coating and the colloidal additivation, while the colloidally additivated specimen shows a slightly higher density than the dry-coated counterpart. These results can be ascribed to the better layer bonding of lower quantities of CNP and the higher dispersion of nanoparticles by the colloidal additivation.

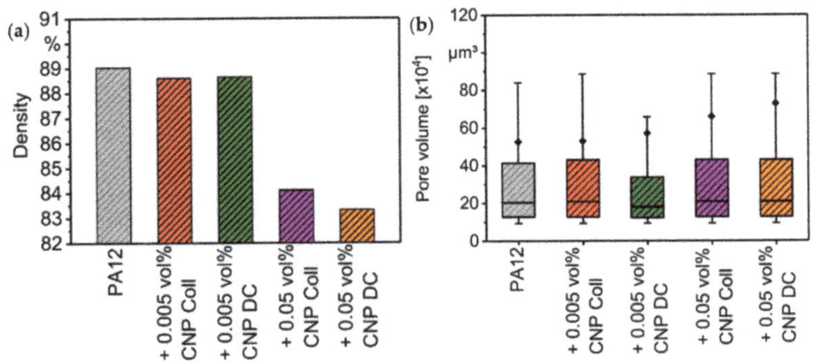

Figure 11. (**a**) Relative density and (**b**) pore size distribution of pure PA12 and PA12 specimens with CNP additivation.

Figure 11b shows the size distribution, average pore size, and the pore counts within each specimen. In most of the specimens, pore volumes lay between 10×10^4 µm^3 and 85×10^4 µm^3. The average pore volume also increases as the quantity of CNP in the PA12 specimens increases. The colloidal additivation of 0.005 vol% CNP reduces the maximum pore size of PA12 from 85×10^4 µm^3 to 69×10^4 µm^3, while the dry coating method and higher amounts of CNP increase the maximum pore size by 5×10^4 µm^3. Therefore, lower quantities of CNP are less detrimental for the density of parts, in which case the dispersion of the nanoparticles plays a crucial role in the final part properties.

Only very small pores in the specimens have high sphericity of up to approximately 0.8; however, it decreases drastically with the increase in the pore volume in all the specimens (Figure 12). As mentioned, large pores are long void spaces between the scan layers that are present due to a lack of fusion, which correlates with the increase in the lack of layer bonding when adding CNP to PA12 (Figure 7). As mentioned before, the process parameters were not chosen to achieve the highest density for every material composition but to investigate the influence of the nano-additivation on the material properties.

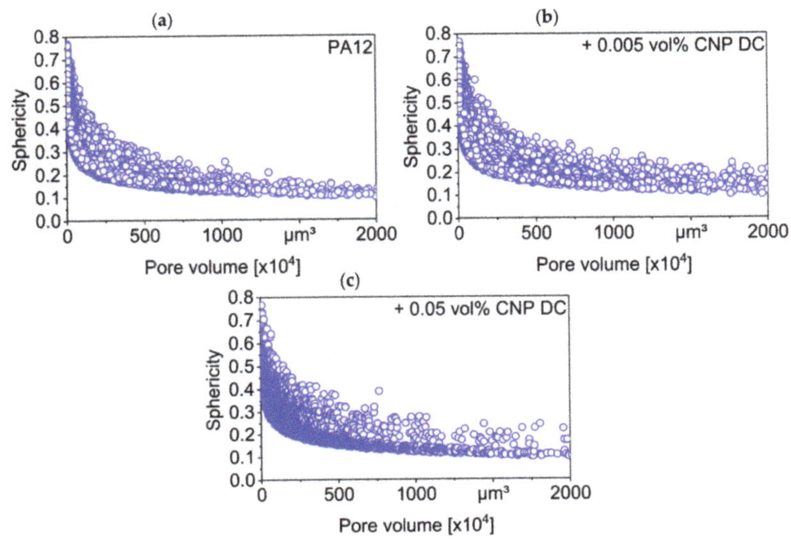

Figure 12. Sphericity measurements of the pores in specimens of (**a**) pure PA12, (**b**) PA12 dry-coated with 0.005 vol% CNP and (**c**) PA12 dry-coated with 0.05 vol% CNP.

4. Conclusions

Well-dispersed nano-additives on polymer powders for laser-based powder bed fusion show potential for the tuning of material properties, but also pose challenges that must be overcome. In our study, we processed polyamide 12 powder that was modified with sub-monolayer quantities of carbon black nanoparticles, by means of two additivation methods of different dispersion qualities, to standardized tensile bars under the same process conditions. The two methods of polymer particle nano-coating were the aqueous colloidal deposition and the dry mechanical mixing.

The differences between the two deposition methods become evident at different stages of the laser-based powder bed fusion process. During the evaluation of the powder flowability, only the colloidal additivation procedure keeps the free-flowing characteristic of pure polyamide 12 powder. This characteristic has an impact on the powder application of the powder bed fusion process, in which mechanically mixed powders lead to a less homogeneous powder bed. After processing the nanocomposites under same process conditions, 0.005 vol% of colloidally deposited carbon nanoparticles show an increase in the ductile material behavior of manufactured parts at the expense of tensile strength, while mechanically admixed nanoparticles reduce the mechanical properties of PA12. Despite increasing the crystallinity, carbon nanoparticles do not change the crystalline morphology of the intermediate form between the alpha and gamma phases of manufactured specimens. Higher amounts of nanoparticles than 0.005 vol% result in poorer layer bonding, reduced part densities with enlarged pores, and thus, worse mechanical properties, regardless of the deposition method. Ultimately, the quality of the dispersion of minute amounts of nanoparticles is critical to tailor the mechanical properties of thermoplastic parts by laser-based powder bed fusion.

Future studies should include optimization of the laser-based powder bed fusion process of polymer nanocomposites towards high part densities of carbon nano-additivated specimens achieved through the adjustment of the powder bed temperature and volume energy density. Furthermore, lasers other than CO_2 (in the near-infrared or visual wavelength range) should be used to facilitate an improved understanding of the influence of absorption-enhancing nanoparticles on the laser–material interaction.

Supplementary Materials: The following are available online at https://www.mdpi.com/article/10.3390/ma14185322/s1, Figure S1: nominal and measured laser power, Table S1: process parameters for LB-PBF-P, Table S2: powder properties, Table S3: d-spacings (from XRD) of the manufactured specimens.

Author Contributions: Conceptualization, A.S., M.S., B.G. and S.B.; methodology, A.S. and S.G.; formal analysis, A.S., J.S., A.B.C. and A.L.; investigation, A.S., S.G., J.S., A.B.C. and A.L.; resources, S.G.; data curation, A.S. and J.S.; writing—original draft preparation, A.S.; writing—review and editing, A.S., S.G., J.S., A.B.C., A.L., B.G., S.B. and M.S.; visualization, A.S., B.G., S.B. and M.S.; supervision, F.W., B.G., S.B. and M.S.; project administration, A.S. and M.S.; funding acquisition, F.W., B.G., S.B. and M.S. All authors have read and agreed to the published version of the manuscript.

Funding: The authors gratefully acknowledge the funding provided by the Erlangen Graduate School in Advanced Optical Technologies (SAOT) through the Bavarian State Ministry for Science and Art. The authors gratefully acknowledge the funding received from the German Research Foundation (DFG) within the priority program (SPP) 2122 "Materials for Additive Manufacturing" (MATframe, WA 1672/53-1, BA 3580/27-1 and SCHM 2115/78-1). B.G. additionally acknowledges funding from the DFG within the Heisenberg Program, project number 445127149 (GO 2566/10-1).

Institutional Review Board Statement: Not applicable.

Informed Consent Statement: Not applicable.

Data Availability Statement: The data presented in this study are available on request from the corresponding author. The data are not publicly available due to privacy.

Acknowledgments: We thank Tim Hupfeld for the delivery of the nano-additivated polyamide 12 powders.

Conflicts of Interest: The authors declare no conflict of interest.

References

1. Bain, E.D. Polymer powder bed fusion additive manufacturing: Recent developments in materials, processes, and applications. In *Polymer-Based Additive Manufacturing: Recent Developments*; ACS Symposium Series; American Chemical Society: Washington, DC, USA, 2019; Volume 1315, pp. 7–36. ISBN 9780841234260.
2. Deja, M.; Zielinski, D. A pilot study to assess an in-process inspection method for small diameter holes produced by direct metal laser sintering. *Rapid Prototyp. J.* **2019**, *26*, 418–436. [CrossRef]
3. Rehman, A.U.; Sglavo, V.M. 3D printing of geopolymer-based concrete for building applications. *Rapid Prototyp. J.* **2020**, *26*, 1783–1788. [CrossRef]
4. Kusoglu, I.; Doñate-Buendía, C.; Barcikowski, S.; Gökce, B. Laser powder bed fusion of polymers: Quantitative research direction indices. *Materials* **2021**, *14*, 1169. [CrossRef]
5. Higa, C.F.; Gradowski, T.; Elifio-Esposito, S.; De Oliveira, M.F.; Inforçatti, P.; Da Silva, J.V.L.; Amorim, F.L.; Meruvia, M.S. Influence of selective laser sintering process parameters on microstructure and physicochemical properties of poly(vinyl alcohol) for the production of scaffolds. *Rapid Prototyp. J.* **2020**, *26*, 1155–1164. [CrossRef]
6. Aldahash, S.A.; Gadelmoula, A.M. Orthotropic properties of cement-filled polyamide 12 manufactured by selective laser sintering. *Rapid Prototyp. J.* **2020**, *26*, 1103–1112. [CrossRef]
7. Bourell, D.; Coholich, J.; Chalancon, A.; Bhat, A. Evaluation of energy density measures and validation for powder bed fusion of polyamide. *CIRP Ann.* **2017**, *66*, 217–220. [CrossRef]
8. Snow, Z.; Martukanitz, R.; Joshi, S. On the development of powder spreadability metrics and feedstock requirements for powder bed fusion additive manufacturing. *Addit. Manuf.* **2019**, *28*, 78–86. [CrossRef]
9. Bonilla, J.S.G.; Dechet, M.A.; Schmidt, J.; Peukert, W.; Bück, A. Thermal rounding of micron-sized polymer particles in a downer reactor: Direct vs indirect heating. *Rapid Prototyp. J.* **2020**, *26*, 1637–1646. [CrossRef]
10. Hupfeld, T.; Salamon, S.; Landers, J.; Sommereyns, A.; Doñate-Buendía, C.; Schmidt, J.; Wende, H.; Schmidt, M.; Barcikowski, S.; Gökce, B. 3D printing of magnetic parts by laser powder bed fusion of iron oxide nanoparticle functionalized polyamide powders. *J. Mater. Chem. C* **2020**, *8*, 12204–12217. [CrossRef]
11. Cai, C.; Tey, W.S.; Chen, J.; Zhu, W.; Liu, X.; Liu, T.; Zhao, L.; Zhou, K. Comparative study on 3D printing of polyamide 12 by selective laser sintering and multi jet fusion. *J. Mater. Process. Technol.* **2020**, *288*, 116882. [CrossRef]
12. Espera, A.H.; Valino, A.D.; Palaganas, J.; Souza, L.; Chen, Q.; Advincula, R.C. 3D printing of a robust polyamide-12-carbon black composite via selective laser sintering: Thermal and electrical conductivity. *Macromol. Mater. Eng.* **2019**, *304*, 1800718. [CrossRef]
13. Athreya, S.R.; Kalaitzidou, K.; Das, S. Processing and characterization of a carbon black-filled electrically conductive Nylon-12 nanocomposite produced by selective laser sintering. *Mater. Sci. Eng. A* **2010**, *527*, 2637–2642. [CrossRef]
14. Athreya, S.R.; Kalaitzidou, K.; Das, S. Mechanical and microstructural properties of Nylon-12/carbon black composites: Selective laser sintering versus melt compounding and injection molding. *Compos. Sci. Technol.* **2011**, *71*, 506–510. [CrossRef]
15. Relinque, J.J.; García-Romero, M.G.; Hernández-Saz, J.; Navas, J.; Gil-Mena, A.J.; Sales, D.L.; Navas, F.J.; Morales-Cid, G.; Aguilera, D.; Periñán, A.; et al. Influence of the additivation of graphene-like materials on the properties of polyamide for Powder Bed Fusion. *Prog. Addit. Manuf.* **2018**, *3*, 233–244. [CrossRef]
16. Yuan, S.; Zheng, Y.; Chua, C.K.; Yan, Q.; Zhou, K. Electrical and thermal conductivities of MWCNT/polymer composites fabricated by selective laser sintering. *Compos. Part A Appl. Sci. Manuf.* **2018**, *105*, 203–213. [CrossRef]
17. Bai, J.; Goodridge, R.D.; Hague, R.J.; Song, M. Improving the mechanical properties of laser-sintered polyamide 12 through incorporation of carbon nanotubes. *Polym. Eng. Sci.* **2013**, *53*, 1937–1946. [CrossRef]
18. Goodridge, R.; Shofner, M.; Hague, R.; McClelland, M.; Schlea, M.; Johnson, R.; Tuck, C. Processing of a Polyamide-12/carbon nanofibre composite by laser sintering. *Polym. Test.* **2011**, *30*, 94–100. [CrossRef]
19. Bai, J.; Goodridge, R.D.; Hague, R.J.; Song, M.; Murakami, H. Nanostructural characterization of carbon nanotubes in laser-sintered polyamide 12 by 3D-TEM. *J. Mater. Res.* **2014**, *29*, 1817–1823. [CrossRef]
20. Hupfeld, T.; Sommereyns, A.; Schuffenhauer, T.; Zhuravlev, E.; Krebs, M.; Gann, S.; Keßler, O.; Schmidt, M.; Gökce, B.; Barcikowski, S. How colloidal surface additivation of polyamide 12 powders with well-dispersed silver nanoparticles influences the crystallization already at low 0.01 vol%. *Addit. Manuf.* **2020**, *36*, 101419. [CrossRef]
21. Doagou-Rad, S.; Islam, A.; Jensen, J.S.; Alnasser, A. Interaction of nanofillers in injection-molded graphene/carbon nanotube reinforced PA66 hybrid nanocomposites. *J. Polym. Eng.* **2018**, *38*, 971–981. [CrossRef]
22. Rahaman, M.; Khastgir, D.; Aldalbahi, A.K. (Eds.) Carbon-containing polymer composites. In *Springer Series on Polymer and Composite Materials*; Springer: Singapore, 2019; ISBN 978-981-13-2687-5.
23. Sommereyns, A.; Hupfeld, T.; Gann, S.; Wang, T.; Wu, C.; Zhuravlev, E.; Lüddecke, A.; Baumann, S.; Rudloff, J.; Lang, M.; et al. Influence of sub-monolayer quantities of carbon nanoparticles on the melting and crystallization behavior of polyamide 12 powders for additive manufacturing. *Mater. Des.* **2021**, *201*, 109487. [CrossRef]

24. Webster, T.J.T. Mechanical properties of dispersed ceramic nanoparticles in polymer composites for orthopedic applications. *Int. J. Nanomed.* **2010**, *5*, 299–313. [CrossRef]
25. Hupfeld, T.; Sommereyns, A.; Riahi, F.; Doñate-Buendía, C.; Gann, S.; Schmidt, M.; Gökce, B.; Barcikowski, S. Analysis of the nanoparticle dispersion and its effect on the crystalline microstructure in carbon-additivated PA12 feedstock material for laser powder bed fusion. *Materials* **2020**, *13*, 3312. [CrossRef]
26. Bain, E.D.; Garboczi, E.J.; Seppala, J.E.; Parker, T.C.; Migler, K.B. AMB2018-04: Benchmark physical property measurements for powder bed fusion additive manufacturing of polyamide 12. *Integr. Mater. Manuf. Innov.* **2019**, *8*, 335–361. [CrossRef]
27. Sommereyns, A.; Hupfeld, T.; Gökce, B.; Barcikowski, S.; Schmidt, M. Evaluation of essential powder properties through complementary particle size analysis methods for laser powder bed fusion of polymers. *Procedia CIRP* **2020**, *94*, 116–121. [CrossRef]
28. Zhang, D.; Gökce, B.; Barcikowski, S. Laser synthesis and processing of colloids: Fundamentals and applications. *Chem. Rev.* **2017**, *117*, 3990–4103. [CrossRef] [PubMed]
29. Hupfeld, T.; Doñate-Buendía, C.; Krause, M.; Sommereyns, A.; Wegner, A.; Sinnemann, T.; Schmidt, M.; Gökce, B.; Barcikowski, S. Scaling up colloidal surface additivation of polymer powders for laser powder bed fusion. *Procedia CIRP* **2020**, *94*, 110–115. [CrossRef]
30. Hupfeld, T.; Laumer, T.; Stichel, T.; Schuffenhauer, T.; Heberle, J.; Schmidt, M.; Barcikowski, S.; Gökce, B. A new approach to coat PA12 powders with laser-generated nanoparticles for selective laser sintering. *Procedia CIRP* **2018**, *74*, 244–248. [CrossRef]
31. Hupfeld, T.; Wegner, A.; Blanke, M.; Doñate-Buendía, C.; Sharov, V.; Nieskens, S.; Piechotta, M.; Giese, M.; Barcikowski, S.; Gökce, B. Plasmonic seasoning: Giving color to desktop laser 3D printed polymers by highly dispersed nanoparticles. *Adv. Opt. Mater.* **2020**, *8*, 2000473. [CrossRef]
32. Chatham, C.A.; Long, T.E.; Williams, C.B. A review of the process physics and material screening methods for polymer powder bed fusion additive manufacturing. *Prog. Polym. Sci.* **2019**, *93*, 68–95. [CrossRef]
33. Lüddecke, A.; Pannitz, O.; Zetzener, H.; Sehrt, J.; Kwade, A. Powder properties and flowability measurements of tailored nanocomposites for powder bed fusion applications. *Mater. Des.* **2021**, *202*, 109536. [CrossRef]
34. Hesse, N.; Winzer, B.; Peukert, W.; Schmidt, J. Towards a generally applicable methodology for the characterization of particle properties relevant to processing in powder bed fusion of polymers—From single particle to bulk solid behavior. *Addit. Manuf.* **2021**, *41*, 101957. [CrossRef]
35. Pigliaru, L.; Rinaldi, M.; Ciccacci, L.; Norman, A.; Rohr, T.; Ghidini, T.; Nanni, F. 3D printing of high performance polymer-bonded PEEK-NdFeB magnetic composite materials. *Funct. Compos. Mater.* **2020**, *1*, 1–17. [CrossRef]
36. Rwei, S.-P.; Ranganathan, P.; Lee, Y.-H. Isothermal crystallization kinetics study of fully aliphatic PA6 copolyamides: Effect of novel long-chain polyamide salt as a comonomer. *Polymers* **2019**, *11*, 472. [CrossRef]
37. Gogolewski, S.; Czerntawska, K.; Gastorek, M. Effect of annealing on thermal properties and crystalline structure of polyamides. Nylon 12 (polylaurolactam). *Colloid Polym. Sci.* **1980**, *258*, 1130–1136. [CrossRef]
38. Jenike, A.W. *Storage and Flow of Solids*; Bulletin No. 123 of the Utah Engineering Experiment Station; The University of Utah: Salt Lake City, UT, USA, 1976; Volume 53, No. 26.
39. Dadbakhsh, S.; Verbelen, L.; Verkinderen, O.; Strobbe, D.; Van Puyvelde, P.; Kruth, J.-P. Effect of PA12 powder reuse on coalescence behaviour and microstructure of SLS parts. *Eur. Polym. J.* **2017**, *92*, 250–262. [CrossRef]
40. Hartley, P.; Parfitt, G.; Pollack, L. The role of the van der Waals force in the agglomeration of powders containing submicron particles. *Powder Technol.* **1985**, *42*, 35–46. [CrossRef]
41. Pinchuk, A. Size-dependent Hamaker constant for silver nanoparticles. *J. Phys. Chem. C* **2012**, *116*, 20099–20102. [CrossRef]
42. Powell, A.W.; Stavrinadis, A.; De Miguel, I.; Konstantatos, G.; Quidant, R. White and brightly colored 3D printing based on resonant photothermal sensitizers. *Nano Lett.* **2018**, *18*, 6660–6664. [CrossRef]
43. Schmid, M.; Kleijnen, R.; Vetterli, M.; Wegener, K. Influence of the origin of polyamide 12 powder on the laser sintering process and laser sintered parts. *Appl. Sci.* **2017**, *7*, 462. [CrossRef]
44. Guo, J.; Bai, J.; Liu, K.; Wei, J. Surface quality improvement of selective laser sintered polyamide 12 by precision grinding and magnetic field-assisted finishing. *Mater. Des.* **2018**, *138*, 39–45. [CrossRef]
45. Cai, Z.; Bao, H.; Zhu, C.; Zhu, S.; Huang, F.; Shi, J.; Hu, J.; Zhou, Q. Structure evolution of polyamide 1212 during the uniaxial stretching process: In situ synchrotron wide-angle X-ray diffraction and small-angle X-ray scattering analysis. *Ind. Eng. Chem. Res.* **2016**, *55*, 7621–7627. [CrossRef]
46. Dai, R.; Huang, M.; Ma, L.; Liu, W.; He, S.; Liu, H.; Zhu, C.; Wang, Y.; Zhang, Z.; Sun, A. Study on crystal structure and phase transitions of polyamide 12 via wide-angle X-ray diffraction with variable temperature. *Adv. Compos. Hybrid Mater.* **2020**, 522–529. [CrossRef]
47. Ma, N.; Liu, W.; Ma, L.; He, S.; Liu, H.; Zhang, Z.; Sun, A.; Huang, M.; Zhu, C. Crystal transition and thermal behavior of Nylon 12. *e-Polymers* **2020**, *20*, 346–352. [CrossRef]
48. Stamhuis, J.; Pennings, A. Crystallization of polyamides under elevated pressure: Pressure-induced crystallization from the melt and annealing of folded-chain crystals of nylon-12, polylaurolactam under pressure. *Polymer* **1977**, *18*, 667–674. [CrossRef]
49. Wang, D.; Shao, C.; Zhao, B.; Bai, L.; Wang, X.; Yan, T.; Li, J.; Pan, G.; Li, L. Deformation-induced phase transitions of polyamide 12 at different temperatures: An in situ wide-angle x-ray scattering study. *Macromolecules* **2010**, *43*, 2406–2412. [CrossRef]
50. Ramesh, C. Crystalline transitions in nylon 12. *Macromolecules* **1999**, *32*, 5704–5706. [CrossRef]

51. Cojazzi, G.; Fichera, A.; Garbuglio, C.; Malta, V.; Zannetti, R. The crystal structure of polylauryllactam (nylon 12). *Die Makromol. Chem.* **1973**, *168*, 289–301. [CrossRef]
52. Rhee, S.; White, J.L. Crystal structure and morphology of biaxially oriented polyamide 12 films. *J. Polym. Sci. Part B Polym. Phys.* **2002**, *40*, 1189–1200. [CrossRef]

MDPI
St. Alban-Anlage 66
4052 Basel
Switzerland
Tel. +41 61 683 77 34
Fax +41 61 302 89 18
www.mdpi.com

Materials Editorial Office
E-mail: materials@mdpi.com
www.mdpi.com/journal/materials